The Ecology of Collective Behavior

The Ecology of Collective Behavior

DEBORAH M. GORDON

PRINCETON UNIVERSITY PRESS
PRINCETON & OXFORD

Published by Princeton University Press
41 William Street, Princeton, New Jersey 08540
99 Banbury Road, Oxford OX2 6JX

press.princeton.edu

ISBN 978-0-691-23214-0
ISBN (pbk.) 978-0-691-23215-7
ISBN (e-book) 978-0-691-23216-4

British Library Cataloging-in-Publication Data is available

Editorial: Alison Kalett and Hallie Schaeffer
Production Editorial: Jill Harris
Cover Design: Heather Hansen
Production: Lauren Reese
Publicity: William Pagdatoon
Copyeditor: Cynthia Buck

Cover images: Cells (top left) by Kevin Cheung, Fred Hutch, and Andy Ewald, Johns Hopkins; ants (top right) by Dennak Murphy; cheetahs (bottom) by Sam Crow

This book has been composed in Arno and Sans

10 9 8 7 6 5 4 3 2 1

For there is no creature whose inward being is so strong that it is not greatly determined by the world around it.

—GEORGE ELIOT, *MIDDLEMARCH*

CONTENTS

The Ecology of Collective Behavior

1

Introduction

Collective behavior is familiar, but difficult to explain. We see it everywhere in nature, and we engage in it ourselves. That you are reading this book depends on many forms of collective behavior, from the language it is in, which we acquire and maintain by using it with each other, to the paper it is on, made out of trees whose growth depends on the intricate relations of their cells. Though it is clear that collective behavior arises from interactions among the participants, it is hard to say exactly how. For example, brains work through the collective behavior of neurons. We know that neurons interact by turning each other on, but how does this produce perception, memory, and even books about collective behavior? The answer is not in the properties of individual neurons. Neurons function in brains in bodies responding to a world teeming with pattern, change, and other beings. It is these relations that generate the connections among neurons that elicit the action of each one.

Whether the whole is more than the sum of the parts is an artificial problem, created by taking the whole apart. All the king's horses and all the king's men could not put Humpty Dumpty together again, but only after he toppled off the wall; while he was alive, the cells in his shell interacted to hold him together.

The key to understanding how collective behavior works is in the dynamic relation of inside and outside rather than the assembly of parts into wholes. We are impeded by the long history that we have inherited of considering a natural system as a bounded package that sits in a separate outside. This view leads us to account for human behavior as the actions of independent beings, propelled by internal decisions and motivations, or to partition the causes of behavior between innate or environmental. But it is not possible to learn how natural systems function collectively by considering the components separately and independently of the world they inhabit. The attempt to parse the internal and external into distinct and additive forces severs the relations that matter.

The approach I describe here has a different starting point: to explain collective behavior, it is necessary to think about how it changes in relation with the changing world it is in. I call this approach "ecological" because ecology (derived from *oikos*, the Greek word for "village") is the study of the interactions that regulate systems. It investigates dynamic systems of relations rather than sets of individuals with independent, internal properties. Versions of this, now sometimes called a "systems approach," flow through the history of the social and natural sciences.

My path to this perspective began in my second year of graduate school, when I was thinking of quitting. I had chosen to go to graduate school to learn how to do research in animal behavior and ecology because that seemed more attractive than the three alternatives I could think of: it entailed less blood than medical school, much more comfortable shoes than law or business school, and more time outside than any of the others. I had only a vague idea of what it would be like to do research. My experience of biology in undergraduate classes was mostly an effort to memorize for exams the names of parts and the little arrows connecting them—gene transcription, the Krebs cycle, photosynthesis—but these diagrams floated disconnected in a blank void in my mind, much like the white background in the textbook illustrations. Then, in my senior year, I took a course in comparative anatomy. It was a revelation to learn that the diversity of body forms in animals reflects their evolutionary history. I had been thrilled to discover, in other classes, the order in human creations, such as the counterpoint of a Bach fugue or a logical proof, but this was even more amazing: evolution generates order in a process that makes itself.

I had imagined that in grad school I would be given a lens through which I could see the processes that regulate nature, so it was baffling at first to experience the opposite. As I learned more about scientific accounts of animal behavior, it seemed that research questions were framed in a way that cut up the world into unrecognizable pieces. Behavior was portrayed as a string of prefabricated and independent snippets, and the task of researchers was to find the external switch that set off the distinct internal mechanism that initiated each snippet, or to explain why it was all for the best that the animal responded that way. An animal was a kind of windup toy, and we were looking for the key that would get it moving. But watching birds, as I did at first, or any other animals, I didn't see windup toys buzzing around. I saw rhythm, pattern, and flexibility—changing behavior linked to the behavior of others and the changing surroundings.

Each week I met with my adviser, John Gregg, and tried to talk about my sense of unease. He just smiled tolerantly and handed me books to read. One of them was Donna Haraway's *Crystal, Fabrics, and Fields,* in which I first encountered a framework and a vocabulary that made it possible to talk about my discomfort.[1] It was a great relief to discover that there was a history of thinking about biology as systems of relations. Haraway's book traces the history of the debates in developmental biology in the late nineteenth and early twentieth centuries, centered on the question of whether each cell in a developing embryo arrives at a function by following inner instructions, independently of the other cells, or instead as a result of its encounters with other cells. The debate was fueled by the opposing camps' choices of study organisms. On one side, Hans Driesch studied sea urchins, whose embryo's cells, if separated, can take on different functions, suggesting that conditions and encounters outside a cell determine its fate. On the other side, Wilhelm Roux studied frogs, in which cell function seemed to be predetermined, presumably by something inside each cell.

It was encouraging to me to learn of Driesch's perspective and the discussions it stimulated among biologists, because this suggested to me a way forward. If we could locate the cause of a cell's behavior in its relations with other cells and the surroundings, we could think of the causes of an animal's behavior in the same way.

I didn't leave grad school. Instead, I began to study ants as a way of investigating these questions in a system where I could see what was happening as it happened. This is difficult to do in a living embryo (though the advent of the confocal microscope has now opened the way[2]). Ants are easy to watch, and even then, I knew that I learn best by seeing. Ant colonies, like embryos, operate without central control; the queen merely lays the eggs. But each one functions only in relation with the others, as Driesch saw with the cells of a sea urchin embryo.

When I began research on ants, they were viewed in scientific work as little automata, each genetically programmed to do its task, triggered by chemical cues. My first research project on ants showed that the response of an ant to a particular chemical depends on the group of ants that it is working with—either the ants collecting food or those bringing out the garbage. Thus, an ant's response depended on its relation with other ants, just as a cell's function depends on its relation with other cells. This observation meant that there was more to understanding ant behavior than identifying a chemical that acted as a switch. I was looking for a different approach.

I found ways to ask research questions that link behavior with what is around it, beginning by observing harvester ants in the desert of the US Southwest. I learned how harvester ant colonies use a network of simple olfactory interactions to regulate activity in response to changing conditions; how the ants of neighboring colonies interact to partition a foraging area; and how all of this changes over the lifetime of a colony. By following a population over generations of colonies, it has become possible to learn how natural selection from water stress is shaping each ant's response to encounters with other ants, and how this will shift as climate change exacerbates an ongoing drought. The alternative to the quest for switches that trigger isolated and fixed pieces of behavior was to follow the layers of relations and responses to changing situations.

Along the way I gave up on talking about parts and wholes, and on trying to distinguish behavior at the individual, group, or colony level, instead asking how collective behavior at each layer is generated by interactions among participants and how this responds to changing conditions. Collaborations with mathematical biologists, computer scientists, and engineers led to quantitative descriptions of how ant colonies work. Such models, not just for ant colonies but for many other systems, show how collective behavior can be explained. We can learn enough about a system to specify how the participants interact, how these interactions are related to the current situation, and how the interactions produce the outcome.

The relation of inside and outside is always dynamic because everything changes—or, as Heraclitus put it in 500 BC (and as discussed more recently by Daniel Nicholson and John Dupre), *panta rhei* (everything flows).[3] Response to changing conditions is fundamental in living systems. Different fields of biology give this different names, such as "kinetics" in molecular and cell biology," "regulation" in physiology, or "adaptation" in neuroscience. An organism's phenotype may seem static—the flower has a certain number of petals, a person has brown or blue eyes—but even apparently static features are the outcome of some process, such as the unfolding of a flower from a bud, or the interactions among cells with melanin in the development of the layers of the iris. However any living entity may appear now, it was not always like that and will not always be the same. Change is built into life on earth, whose rotation generates the cycles of night, day, and the pull of the tides, setting up daily patterns of movement, growth, feeding, and rest. The planet's annual journey around the sun brings seasons, rain and sun, heat and cold. On a smaller scale, gut bacteria ride waves of peristalsis; blood pushes through a tangle of bifurcating tubes with each heartbeat; and molecules jiggle around causing proteins to shift their attachments.

In living systems, unlike physical ones, change goes both ways: change in any living system alters its surroundings, which in turn change the living system.[4] Newton's laws describe how an object changes position when it is subject to certain forces. These laws for inanimate objects are deterministic; the relation of mass and acceleration is sufficient to predict how an object will move, and the surroundings are independent of the action of the objects.[5] This is never true in living systems; instead, every living entity is busy modifying its surroundings. This is easy to see in plants and animals, but equally important at every scale. For example, breast cancer cells change how neighboring cells interact with the surrounding collagen matrix and facilitate their own movement and metastasis by reorganizing the spatial pattern of cells.[6] Such mutual modification of living systems and their surroundings ripples across species. As a hummingbird flies from flower to flower and dips its beak to collect nectar, it carries microbes from the nectar of one flower to the next flower, thus changing the community of microbes living in the flowers, which in turn modify the nectar that the hummingbird eats.[7]

Here I propose that the dynamics of environments provide clues to how collective behavior operates. There are likely to be trends in how interactions generate collective outcomes, according to how the conditions in which the behavior functions are changing. The kinetics of biochemical reactions within and between cells, the regulation of physiological processes in tissues, the behavior of plants and animals in particular habitats, all reflect the way that their surroundings change. Comparative approaches in evolutionary biology show that analogous traits have evolved in relation with similar environments. Because new phenotypes tend to arise as effective responses to changing surroundings, and because adaptation leads these new phenotypes to persist over the course of evolution, similar innovations in the regulation of collective behavior should arise in conditions that change in similar ways. For example, mammals that live in cold places are likely to have thicker fur than those that live in hot places, because thick fur keeps an animal warm when it's cold. By analogy, collective behavior that responds to rapidly changing conditions is likely to use interactions that can adjust rapidly.

In *Evolution in Changing Environments,* Richard Levins outlines a general principle for explaining the evolution of phenotypic plasticity, the capacity to change in response to changing situations.[8] He argues that there is a trade-off between two costs. The first is the cost of the work it takes to be plastic, so as to have the capacity to change when the situation changes, rather than just staying the same. The second is the cost of having a wrong or inadequate response in a particular situation. In Levins's model, the capacity to change evolves when the cost of having the wrong response in a situation is greater than the cost of the capacity to change.

This idea can be applied to the evolution of plasticity of collective behavior, the capacity to regulate or adjust collectively to the current situation. Regulation evolves when it's important to have the right response in a particular condition. Or, to turn this around, regulation that adjusts appropriately to conditions is likely to evolve. Thus, understanding the conditions to which a form of collective behavior responds suggests hypotheses about how the behavior works. For example, adrenaline influences the response to sudden change. If we knew nothing about the physiology of the adrenal system, we would guess that the processes that regulate adrenaline involve rapid chemical interactions, because the events in the surroundings that are relevant to adrenaline—such as danger—require a rapid response.

To outline correspondences between the dynamics of collective behavior and of its surroundings, I choose some features of each and suggest how they are associated. To characterize the dynamics of collective behavior—how interactions respond to changing conditions—I draw on ideas from dynamical systems and network science. I consider first, the rate at which interactions adjust to conditions, and second, the feedback regime that stimulates and inhibits a collective process. Third, I consider how participants in a collective process are linked in a network of interactions.

To characterize the dynamics of environments, I use ideas from ecology that describe three gradients in changing conditions. The first gradient is stability: how frequently and how much conditions change. This gradient includes the risk or probability of a rupture or adverse event. The second gradient is the distribution in space and time of the resources that the system uses or the needs and demands that the system must respond to. A simple version of this gradient goes from scattered or random to clustered or patchy. The third gradient is in the energy flow that conditions allow, from a high to low ratio of energy or resources used to energy or resources taken in.

To raise new questions about collective behavior from an ecological perspective, I propose three main hypotheses, all of which suggest that, across different natural systems, similar kinds of changing conditions correspond to similar ways of using interactions to regulate collective behavior. First, the rate at which interactions adjust collective behavior is associated with the stability of the environment and the distribution of resources and demands. Second, the feedback that regulates interactions is associated both with the distribution of resources and demands and with the energy flow required to operate in that environment. Third, the modularity of interaction networks is associated with stability and the distribution of resources and demands.

To begin, the next chapter introduces these hypotheses briefly by comparing the collective behavior that regulates foraging in two ant species living in very different environments. In chapters 3 and 4, I first define collective behavior very broadly and then discuss quantitative models for how collective behavior arises from interactions among participants. In chapters 5 and 6, the ecological hypotheses are outlined in detail, with examples. Finally, chapters 7, 8, and 9 discuss the research program suggested by these hypotheses, situate this approach in current evolutionary biology, and contrast it with the prevailing one that is based on the idea that collective behavior evolves out of conflict between the interests of the individual and those of the group.

2

The Ecology of Collective Behavior in Ants

There are more than fifteen thousand species of ants, making a living, in different ways, in every terrestrial habitat. Their diversity, in both behavior and ecology, offers an opportunity to learn how collective behavior evolves in relation with changing environments. Of the many species of ants, only about fifty have been studied in detail, but even that is enough to show trends in the relation between collective behavior and changing environments. In this chapter, I illustrate these trends by comparing two species that live in very different surroundings—the red harvester ant (*Pogonomyrmex barbatus*) in the desert of the southwestern United States and the turtle ant (*Cephalotes goniodontus*) in the tropical forest of western Mexico.

Differences in the collective behavior of these two species illustrate the hypotheses proposed here for how the dynamics of collective behavior are linked to the dynamics of its environment.[1] These hypotheses describe how interactions are likely to work to regulate collective behavior in particular environments. How quickly the interactions shift as conditions change, how they stimulate or inhibit activity, and which participants tend to interact with others are all associated with the surroundings.

To ask how interactions are used in different environments to adjust behavior to conditions, I consider: the rate at which interactions respond to conditions, the feedback regime used to activate or inhibit the behavior, and the modularity of the network of interactions. Rate is simply how quickly interactions change in response to conditions. Feedback occurs when the outcome of interactions has a further effect; here I distinguish a feedback regime in which activity is stopped unless it is stimulated from one in which activity continues unless it is inhibited. The modularity of the interaction network is the extent to which interactions tend to occur within subgroups or modules of participants.

To characterize the dynamics of environments I discuss three gradients: stability, which is how often and how much change occurs; the distributions in space and time of the resources that the system uses and of the needs or demands that it responds to; and energy flow, or the ratio of the energy that is used to the energy obtained. These terms and the hypotheses are described in detail in chapters 5 and 6.

The three main hypotheses about correspondences between collective behavior and its environment, illustrated in figure 2.1, are:

1. The rate of interaction is related to stability and the distribution of resources or demands. This rate corresponds to the rate of change in the surroundings to which the behavior responds: the more stable the conditions, the more slowly interactions are adjusted in response to them; the more unstable the conditions, the more rapidly interactions are adjusted. The rate at which interactions respond to the environment also depends on the distribution of resources or demands: the more clustered these are in time or space, the more rapid the adjustment to allow for immediate response.
2. The feedback regime is related to gradients in the stability of conditions and in energy flow. In stable conditions that change slowly, or where energy use is high relative to the energy brought in, the system waits for the opportunity or necessity to act. Positive feedback, where interactions from some activity trigger further activity, can be used when the default is set to inactivity; action is stopped unless something positive

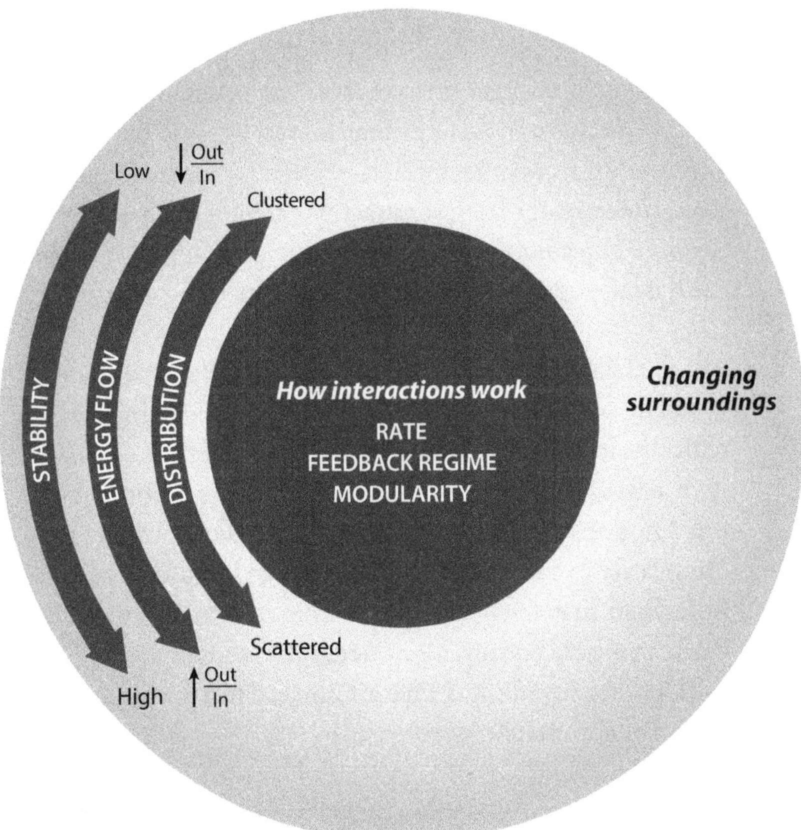

FIGURE 2.1. Framework for investigating the ecology of collective behavior. Rate, feedback regime, and modularity describe how interactions work to adjust collective behavior. This is situated in a changing environment, represented by the outer circle. The three arrows represent three gradients that describe how the surroundings change. The gradient in stability is from high, when change is infrequent or low in magnitude, to low, when change is frequent or high in magnitude. The gradient for energy flow is from a high ratio of energy or resources spent or out to those obtained or in, represented by an upward arrow next to the ratio Out/In, to a low ratio of energy or resources spent or out to those obtained or in, represented by a downward arrow next to the ratio Out/In. The gradient for the distribution in time or space of resources or demands is from scattered or random to patchy or clustered.

happens, thus saving energy unless action is worthwhile. By contrast, in unstable conditions that change frequently, requiring that the system make rapid responses, or when the activity gathers more energy than it uses, the system keeps going unless something inhibits it. Negative feedback, in which interactions from some activity inhibit further activity, can speed up adjustment to rapidly changing conditions when the default is set to keep going unless something negative stops it.

3. The modularity of the interaction network is related to gradients in stability and the distribution of resources. High modularity, with interactions among subgroups of participants, is associated with a rapid response to variable or unstable situations and a rapid response to clustered resources. Networks with high modularity can work faster, be cheaper to build, and have fewer links than networks with low modularity, in which all participants are equally connected. Low modularity can be used in stable conditions where it can lead to a more thorough spread of information or resources.

A comparison of harvester ants and turtle ants illustrates the association between how each species uses interactions to regulate foraging activity and the dynamics of its environment. In an ant colony, the interactions are mostly olfactory and tactile. Ants of most species cannot see, but all of them rely on smell, using their capacity to distinguish hundreds of odors.

Harvester ants live in the desert. The surroundings and the availability of resources change slowly, and resources are scattered, not distributed in patches or clusters. Water is a scarce resource, and the ratio of resources expended to those obtained is high. Since I was a graduate student in the 1980s, I have followed a population of colonies of harvester ants at a site in desert chapparal in New Mexico, near the US-Mexico border. Turtle ants, by contrast, live in the tropical forest, foraging for patchy and ephemeral resources in frequently changing conditions. Activity is easy, so the ratio of energy expended to energy obtained is low. For the past twelve years, I have been studying the trail networks of arboreal turtle ants. The ants nest and forage in the tangled network of vines and branches in the canopy of the forest. After all the time I have spent looking down at harvester ants on the ground in the desert, it is an interesting change of perspective—and good for my neck—to observe ants up in the trees.

A harvester ant colony adjusts its foraging activity from day to day, and hour to hour, in response to changes in humidity and food availability. The interactions used to regulate foraging are brief antennal contacts when one ant touches another's antennae or body with its antenna. Ants smell with their antennae, and during an antennal contact one ant assesses the odor of the other ant. The odors are carried in a layer of grease, cuticular hydrocarbons, that in ants, as in many insects, help to block water loss. Ants spread cuticular hydrocarbons from glands in their mouthparts by grooming, on both their own bodies and those of other ants. Ants of different colonies and of different task groups within a colony smell different because they differ in their cuticular hydrocarbon profile.

Harvester ants, like many other ant species, respond to the rate of antennal contact.[2] Each encounter is like a blank email message from an unnamed sender. It is the pattern, not the content, of the encounters that matters. In other words, the rate of interaction itself is the message, rather than some other message or instruction packed within the interaction. We know this from experiments, done in collaboration with Michael Greene, in which we extracted cuticular hydrocarbons from harvester ants and put the odors on glass beads.[3] Ants responded to the rate at which they encountered and smelled the glass beads. Since the glass bead does not carry any other message besides its odor, the ants must be using just the rate at which they encountered a particular odor.

Harvester ants use antennal contact between outgoing and returning foragers to adjust their foraging activity in response to humidity and to the availability of food. A forager loses water to evaporation while out foraging, but colonies get water from metabolizing the fats in the seeds they eat, so colonies face a trade-off between spending water while searching for food and obtaining food and water. The trade-off is managed using positive feedback. A forager does not leave the nest until it meets enough foragers returning back to the nest with food. Encounters inside the nest between outgoing and returning foragers lead more foragers to leave the nest. Because each forager searches until it finds food, this positive feedback from returning foragers links foraging activity to the current availability of food: the more food is available, the shorter the search time, the more quickly foragers return, and the more foragers go out to search.

The desert environment of harvester ants is stable with respect to resources: the availability of the seeds of annual plants, their main food source, changes slowly. Seeds are distributed mostly by wind and flooding, so seed availability changes on a timescale of weeks and months, much longer than the hour-to-hour regulation of foraging activity. The seeds stay where they are on the ground unless something else eats them; if nothing does, they eventually get buried in the soil. The soil that the ants travel on and the vegetation that they travel past both tend to remain similar within a season, and even from year to year. I once found a pen on the ground at my harvester ant study site that had been there for at least ten years (and it still worked). The availability of water also changes slowly on a seasonal timescale, with rain only in the winter and summer. Colonies compete for foraging area with neighbors of the same species, and the configuration of the neighborhood can persist for decades, as colonies can live for twenty to thirty years.[4]

The collective regulation of foraging activity in harvester ants illustrates the first two hypotheses. The first is that the rate of interaction adjusts slowly in stable environments where resources are scattered and random rather than clustered. The rate of antennal contact between outgoing and returning foragers adjusts slowly to conditions. The rate at which foraging rates can change through the interaction of outgoing and returning foragers at the nest is set by the duration of a foraging trip, which lasts about twenty minutes. The overall rate at which foragers are returning can shift only gradually, and any change in food availability or humidity registers slowly. The regulation of foraging does not use any local spatial information; a forager coming in from one direction can stimulate another to leave in a different direction. Because the resources are scattered in space and time, there is no reason to mobilize a rapid response to the discovery of an ephemeral or concentrated resource. Thus, harvester ant foraging activity does not change rapidly or often, and the magnitude of change is low.

The regulation of harvester ant foraging activity is also consistent with the second hypothesis that the feedback regime is associated with the stability of conditions and the energy flow required to operate. In the feedback regime that regulates foraging in harvester ants, the default is set to remain inactive unless it is worthwhile to forage. Foraging activity, which spends water in evaporation from searching foragers, is suppressed unless there is enough food available to make it worthwhile given the current humidity. Since the food supply changes slowly and colonies store food, activity is paused until foraging is warranted. A forager does not leave the nest on its next trip unless it is stimulated by a high rate of return of foragers with food. A default set at inactivity corresponds to an environment in which energy flow leads to high expenditure relative to what is gained: especially on a dry day, the colony must spend considerable water to get water and food. Whether a forager goes out depends both on the positive feedback linked to food availability and on the negative influence of low humidity: a forager is less likely to go out on its next trip when it experienced low humidity on its last trip. (The influence of humidity is not due to feedback, since the humidity outside the nest does not depend on the rate of forager return.)

In summary, the harvester ants adjust foraging activity slowly, using a centralized interaction network with low modularity and a feedback regime that reduces foraging in unfavorable conditions. This is consistent with a stable environment with high costs of activity.

The very different dynamics of the collective regulation of foraging in turtle ants reflect the dynamics of the tropical forest. In the canopy of the tropical forest. a turtle ant colony maintains a rapidly changing network of trails that links all of the colony's current nests in an ongoing highway circuit. Trails off the new highway may also lead to new nests; turtle ants nest in tunnels created by beetle larvae in dead branches, so colonies must find their nests rather than build them. Exits off the main circuit create temporary trails to food sources, such as flowers that provide nectar. The trails are maintained and adjusted on a timescale of minutes and hours. (How a colony sets up its network of trails each day is a different searching process.) The ants on a trail interact using a volatile trail pheromone that decays quickly by evaporation. These ants deposit pheromone in tiny drops from their abdomen as they walk, and they do this constantly, not just when they have found a food source. At a junction where an ant has a choice of paths, it is likely to follow the one that has been most reinforced by the accumulated pheromone of ants that recently went that way. Thus, the pheromone is the means of interaction among ants, with a lag bounded by the rate of pheromone decay.

Turtle ants in the tropical forest live in an unstable environment. Everything in the tropical forest changes quickly; if you drop a pen, a thick mat of vines will grow over it, making it impossible to find it in a day or two, let alone ten years. The vegetation in which the ants travel changes frequently and unpredictably as they move through junctions of plant stems that move in the wind from moment to moment, change shape from hour to hour as water pressure in the stems changes, grow larger from day to day, and are easily broken by lizards and birds moving through the canopy. Their nests are in dead branches that rot and break off, sending the nest to the ground and the ants pouring back up the trees to find the trail network.

Activity is easy for the turtle ants because the air is humid and walking around is cheap, so the energy spent is low compared to the energy obtained. The distribution of resources is clustered in both space and time. Clustering in space is a consequence of the diversity of the tropical forest, where the same species are rarely neighbors, so a tree with flowers and nectar is unlikely to be near other trees of the same species that flower at exactly the same time. Flowers are clustered in time because they appear suddenly and disappear quickly, and they might at any time be taken over by a competing ant species also foraging for nectar.

In this unstable environment, turtle ants maintain and repair the trail network through interactions at each junction. Local interactions allow for rapid adjustment and for collective search for new food sources and nest sites. Although at junctions in the vegetation ants tend to choose the path that has the most pheromone, they cannot always take the most reinforced path, and the paths cannot stay the same, or the colony would never find new food sources or nest sites. With some small probability, an ant is likely to take a path that is not the most strongly reinforced. This could be considered making a mistake, but it is also exploration, in the same way that a mutation can allow for innovation in evolution.[5] Usually an exploring or befuddled ant that takes a new path off the common trail turns back as soon as it reaches the next junction where other ants have not traveled recently, so that none of the possible paths forward from that junction have any pheromone. But sometimes the ant does not turn back, continues on, and may reach a new food source or potential nest. When enough ants reach it and come back, depositing pheromone as they go, they reinforce a new path to the food or nest.

Local exploration also allows the ants to repair ruptures in the trail, such as when the wind separates plants at a junction, or a passing animal breaks a branch that the ants are using. When this happens, many ants go back and forth to the dead end of the trail, but some ants continue to explore, trying other branches from the junction nearest to the broken branch. This new trail is reinforced by ants that find the same path from the other direction, and eventually a new segment of trail connects the two broken ends of the circuit.

The trail networks of turtle ants show a different feedback regime from the one used to regulate foraging in harvester ants. This difference is associated with differences in the energy flow required to operate in the desert or the tropical forest. In the feedback regime of the turtle ants' trail network, the default is set to keep going unless activity is inhibited. Once established, a trail is difficult to stop. The default is to keep going along the trail unless there is negative feedback, which can rapidly stop the trail. Feedback that sets the default to keep going is associated with an environment where unstable conditions may require a rapid response, or where activity is easy. Humidity in the tropical forest is high, so there is little cost in water for the amount of food brought in. A turtle ant can stay out on the trail for long intervals, unlike harvester ants, which can make only short trips in the dry desert air before they must return to the more humid nest. The turtle ants, by moving along the trail, keep reinforcing the trail and keep the network going, so that the colony remains connected and exploration continues. The flow of the trail slows down only when ants of a competing or predatory species show up, or one of their junctions is broken, as when the two sides are separated in the wind.

The foraging activity of turtle ants can also generate negative feedback that inhibits the rate of flow of ants on the trail. When turtle ant foragers return to the nest, they must unload their liquid food by trophallaxis with other ants at the nest. Delays can result when the rate of forager return becomes high enough to overwhelm the available recipients, and these delays can then slow the rate at which foragers leave the nest to go out on the trail again and continue reinforcing it with pheromone.

Thus, turtle ants illustrate the other side of the first two hypotheses. First, in an unstable or frequently changing environment, turtle ants reinforce or change their trails rapidly by passing through junctions depositing trail pheromone. Second, in an environment that changes frequently, in which activity is easy, the feedback regime is set with the default to keep foraging. The turtle ants continue along the trail unless something stops them.

The contrast between turtle ants and harvester ants in the regulation of foraging also illustrates the third hypothesis: the modularity of interaction networks is related to both stability and the distribution of resources.

Modularity is low in the harvester ant system, which centralizes the transfer of information and regulates foraging activity very slowly. Foraging activity depends on antennal contacts between returning and outgoing foragers at a central nest. An ant must go out on a foraging trip, search, find a seed, bring it back to the nest, and encounter other ants inside the nest before its behavior has any impact on whether another ant goes out to forage.

The low modularity of the interaction network that regulates foraging in harvester ants fits a food supply that does not change rapidly. Not only do the seed resources used by harvester ants change slowly, but they also do not change much in what is needed to deal with them: all of the seed species that the ants encounter require about the same foraging response. When change is rare, such as in the seeds available to a harvester ant colony each day, responses that are slow to adjust are sufficient. The modularity of the task allocation related to foraging is also low in harvester ants. Ants from other task groups are easily recruited to become foragers when extra food becomes available; thus, ants can easily flow into foraging from elsewhere, and there are no modules of ants designated to become foragers.[6]

The turtle ant system, which adjusts rapidly to a rapidly changing environment, has high modularity. The trail is regulated through local interactions among the ants near each node. Keeping the ants together on the trail is prioritized over finding the path that covers the shortest distance.[7] The process that maintains trails tends to find paths with the fewest possible junctions, thus minimizing the number of paths that ants could take and decreasing the opportunities for ants to wander off the trail and get lost. The system keeps the ants together on the trail network.

The searching behavior of the turtle ants reflects the highly modular distribution of their food resources. The ants forage for nectar and flowers that emerge in the sunlight, at the tops and edges of the canopy, and they also collect other resources such as lichens and fungi, which grow in patches. The ants use breadth-first search: in response to a rupture in the trail, they go to the next nearest junction and try that one, continuing that method until there is a path in both directions connecting the broken ends of the trail.[8] When they find a new food source, they also use breadth-first search, traveling around all the closest links. This modular shape in their searching method contributes to keeping the ants close to the shared trail. It also aids in finding food, because their food sources are in modular vegetation: trees have trunks, then large branches, then flowers and leaves that come off of many stems on a few branches. By searching at nearby nodes, the ants that find one food source, such as a flower with nectar, are likely to find more, because flowers are likely to be closely connected to branches that bear other flowers.

Turtle ant colonies thus regulate foraging using a rapidly adjusted, highly modular system with feedback set to keep going as the default. This allows them to shift their networks in response to frequent changes in food supply, vegetation, and the behavior of other species that also change their nest sites and trails frequently. Their collective regulation of foraging allows for rapid response that keeps the network going and maintains its coherence. Ruptures in the trail, which are frequent and unpredictable, can disconnect the nests, leaving some of the larvae without a food supply, so the capacity for rapid repair is important. Trails form to new food sources only if the rate of flow on the trail is high enough to generate recruitment off the main trail and onto the one made by the ants that found the food. Ants of competing species often find the same food sources that turtle ants do, so a process that leads to the formation of alternative trails allows the turtle ants to avoid conflict.

These differences in the regulation of foraging in harvester ants and turtle ants illustrate how collective behavior reflects the surroundings in which it functions. Collective behavior may be especially conspicuous in ants, but it is happening everywhere, at every level of biological organization: among the many structures and entities inside a cell, and among cells within tissues, tissues within organs, and innumerable kinds of groupings of organisms. The next chapter defines collective behavior broadly and describes two of its forms that show up in many natural systems.

3

Collective Behavior

All it takes to see a living entity engaged with others in collective behavior is to zoom out. Everything alive participates in a network of interactions. This is familiar in our own experience. I like Henry James's formulation: "Everyone with whom one had relations had other relations too, so that even optimism was a mixture and peace an embroilment. The only solution was to let everything be embroiled but one's temper and everything spoiled but one's work."[1] Our tempers can get embroiled because we develop specific relations with particular individuals, as do many other animals. But for most natural systems, the interactions do not build relationships based on identity; an ant does not care if it meets ant number 7 or ant number 452; a cell does not prefer to cuddle up to fibroblast A rather than fibroblast B, and the active site on an enzyme does not mind which molecule its electrons come from. Even when individual identity doesn't matter and tempers are not involved, all life is enmeshed in a tangle of relations. Collective behavior is the playing out of these relations.

I use "interaction" broadly to mean any mutually consequential encounter of participants in a system.[2] This generalizes over many different means of interaction, such as chemical signals, electrical stimuli, exchange of nutrients among symbionts, or behavioral displays, and over many different patterns of interaction, such as a signaling pathway or a social network. The participants could be any entities in a living system, from molecules that act as signal transducers in cells to elephants in a herd.

This broad use of "interaction" also bypasses a distinction, used in some fields to define types of interactions, that contrasts two ways in which participants influence each other. The effects of an interaction between two participants can be symmetrical (the same for both), as when two ants touch antennae and both smell the odors on the other's body, or asymmetrical, as when one animal eats another. In networks of neurons in brains, by analogy with computers, interactions are described as the transfer and processing of information from one neuron to another, coded in whether a bit is on or off, or a neuron fires or not. In physiological systems, by analogy with mechanical systems, interactions are classified as sensing or activating, according to whether a participant takes in input from others or triggers some action by others. In ecology, interactions between species are classified by how they change the numbers of one or both species. For example, the interaction of predators and prey species decreases the population numbers of prey, because some get eaten, and increases the population numbers of predators, who use their food to reproduce. All of these forms of interaction, symmetrical or not, produce collective behavior.

Interactions in nature happen through an infinite variety of mechanisms. Animals use vision, sound, smell, and touch. Plants send chemical and tactile signals along plasmodesmata, the membrane-lined channels that link neighboring cells.[3] Neurons use electrical stimulation and the transfer of neurotransmitters. Many kinds of cells interact by means of chemicals, some of which are secreted by others; these chemicals attach to receptors on the surfaces and interior of cells and also interact by contact, pushing on or making room for each other as they grow and move. Bacteria change form and function in response to chemicals secreted into the medium by other bacteria. In biochemical systems, molecules interact by chemical binding and release. The outcomes of all of these interactions are collective behavior.

Collective Outcomes

Whether we choose to investigate behavior as collective is a matter of perspective. An example is the territorial behavior of birds. Julian Huxley described it as a collection of rubber balls that compress with pressure from neighbors and expand when there is more space.[4] This image describes the collective outcome of interactions among neighboring birds. The same behavior could be investigated by asking questions about individuals, such as what features of a male bird's size or color are associated with the size of its territory.

The most familiar examples of collective behavior are found in animal groups. Mammals that live all their lives in groups include all primates, lions, elephants, badgers, wild dogs, meerkats, javelinas, coatimundis, feral domestic cats, dolphins, and seals, to list just a few. These create long-lived social networks that persist from one generation to another and are associated with living and sleeping places, foraging sites, and joint traveling expeditions. Some species of birds, such as cliff swallows and ravens, create similar groups, while mixed-species bird flocks forage in fluid, temporary groups.[5] The social insects, including bees, ants, and termites, live and reproduce together as colonies. Corals and sponges are colonies of individual animals that are physically connected to each other and mutually dependent.

Although some animals are considered to be solitary rather than group-living, they can also be considered to engage in collective behavior. Even the state of aloneness is itself a relation with others. We know this from our own behavior. People who are alone are not removed from the social fabric; avoidance is a form of relation, and the use of language, even in solitude, is a social activity. The same is true for animals, who sometimes avoid each other. Leopards are solitary, except that often they are not—when they are caring for their young, or when they are mating. A leopard might kill an antelope on her own and then travel on her own to find her cub, but then she brings the cub back so they can eat together. Newts are solitary, but they find each other to travel together back and forth from egg-laying sites. Rattlesnakes live alone part of the year but return to group dens for part of the year for mating and resting. Cheetahs are considered solitary, but cheetah cubs hang out together when they are young. As they grow older, they position themselves farther away from each other, and that positioning is itself a social process.

Only habit leads us to restrict collective behavior to interactions within species. The evolution of plants, animals, and fungi began when different species of bacteria joined together to form eukaryotic cells.[6] Every organism is engaged in intimate, ongoing relations with microbes, what Thomas Bosch and Michael Hadfield call "cellular dialogues."[7] More than half of the cells in a person's body are bacteria of thousands of different species, and no one can live without these microbial communities.[8] As Scott Gilbert and his colleagues point out, there may not be any biological individuals of a single species; it seems that all species are intertwined in mutualistic interactions, so that each individual is a holobiont combining actors of many different species.[9] To give two examples, the larvae of marine polychaetes begin to develop only when they contact the mats of certain bacteria,[10] and sea anemones live in partnership with zooxanthellae, which use photosynthesis to make food and regulate function for both partners as ocean temperatures rise.[11] More generally, different species encounter and respond to each other, at sometimes surprising intersections. Recent videos posted online show a crow that seems to be helping a hedgehog across a busy road[12]; a honey badger chasing off several lions many times its size[13]; a badger and jackal working together to defeat a python[14]; and a coyote and badger traveling together to hunt.[15]

The outcome of collective behavior can be to achieve a particular action, such as when a bird flock splits to avoid a predator. But collective behavior can also function to maintain the coherence of a network, such as the calls of howler monkeys that tell the others they are there. The growing field of "collective intelligence" considers ways to evaluate how well a group accomplishes a collective outcome, and how the group's effectiveness depends on participants' handling of errors in perceiving interactions and in executing a response. The original studies were conducted on human team performance in work settings to learn how performance could be improved. For example, a team's capacity to solve problems is not a simple aggregate of the intelligence or competence of its members, but instead depends on the particular ways that they work together.[16] The same perspective can also be used to evaluate how well a collective problem is solved in an engineered system, such as a turbine farm or an air traffic control system.[17]

The questions raised about collective intelligence in human teams or in engineered systems can be extended to natural systems.[18] Asking whether a collective outcome solves a problem effectively is an interesting question. But for many forms of collective behavior in nature, we are not ready to ask that question. Either we do not know enough about how interactions mesh together to regulate how a task is performed, or, although we can identify the interactions, we do not yet understand how they produce a functional outcome. For example, we are not yet ready to evaluate how much better one animal's neurons are than another's at producing memory because we do not know exactly how neurons work together to accomplish this; though we can map physical connections among neurons, we often do not know what those connections do. To measure the effectiveness of interactions at achieving a collective goal, we first need to know how the process works and how it shifts in relation with changing situations. Here I focus on learning how collective behavior works rather than on how well it performs a task.

Collective behavior is extremely diverse. To give some sense of the range of behavior that can be thought of as collective, I consider two general forms that are widespread and well studied. The first, the most familiar form of collective behavior, involves spatial patterns of movement. The second is the regulation of activity that determines which participant performs which function and when.

Movement and Spacing

It is easy to see changing arrangements of individuals in space as collective, in part because we are so used to participating in such patterns ourselves. People regulate many spatial patterns of movement collectively. School for young children is largely devoted to training us to form spatial patterns, such as to sit still at desks in rows facing the teacher, or to move in a group from one place to another. On a trip to Kyoto, I watched a group of children and their teachers on a school trip in a park. A lot of work was devoted to getting the children to stand in a straight line. The teachers moved along the line gently adjusting the position of the children to keep the line perfectly straight. Apparently foreigners were known to be deficient at this skill; at the airport, uniformed attendants reminded errant travelers waiting for a bus to form a line that stayed within a narrow rectangle marked by yellow lines.

Our collective movement patterns allow us to sit around a table sharing a meal or to stream onto a bus. We form lanes when walking on crowded urban sidewalks and in subway stations, places where maintaining a common speed is essential; people slowing down to look at their phones create chaos.[19] In an elevator, people move in a predictable way: the first person goes to the middle, then moves to a corner when the second enters, so the third person goes to another corner, and so on.[20] We can all engage in myriad forms of collective movement.

Human collective movement can have catastrophic results. In Mina, Mecca, in Saudi Arabia in 2015, people were packed into a spiral line waiting to get to the Black Stone, the Hajr Al Aswad. A disturbance caused a stampede, leading the spiral to tighten, and thousands of people were killed. Similarly, a blocked passageway interfered with collective movement at a large gathering in Meron, Israel, in 2021, creating a deadly stampede.

Animals and cells both show many spectacular forms of synchronous movement. Groups of migrating geese form a V when they travel. It looks as though the goose in front, at the tip of the V, is leading the others, but leadership shifts among geese.[21] The V is shaped by local interactions between the geese as they see their neighbors and feel the airflow created by their flight, rather than by anyone following the leader. If something happens to disrupt the status quo, all the birds adjust their positions until the range of distances between them once again creates a V.

Fish schools use interactions based on both vision and the flow of water created by the movement of other fish.[22] Each fish maintains a certain angle relative to other fish around it. In response to an event in the surroundings, such as one or more fish reacting to a predator coming toward them, the others adjust around the one that changed its direction of movement. This causes the school to turn. Other reactions lead to different configurations of the fish: if the radius within which the fish can see each other changes, the shape of the whole school can be transformed from a blob into a ring of circling fish.

Spatial patterns of movement in cells also depend on interactions.[23] For example, groups of neural crest cells move together during the development of a mouse embryo.[24] The cells at the back contract, in response to the expression of E-cadherin, and rearrange themselves into a denser pattern, which creates a wave that pushes the cells in front along.

Collective spatial patterns are ubiquitous in nature. Bacteria form biofilms, which are mats that carpet a surface. Skin cells draw toward the edges of a wound to close it. A swarm of male midges creates a ball that makes them all conspicuous to females. Ant colonies build elaborate networks of trails to connect nests and food sources. Clouds of locusts travel together. A honeybee colony reproduces by forming a group of workers that surround a queen and then leave the nest to fly to a new one. Dolphins collaborate with human fishermen, moving together to drive fish into the nets.[25] A tree adjusts its growth in relation with its neighbors, which influence its access to light, so the shape of each tree in a forest is the outcome of a collective process.[26] Over many generations of trees, shade from large trees inhibits the growth of smaller ones; this collective demographic process is "self-thinning."[27]

Herds of different ungulate species in the African savanna, including antelope, impala, kudu, zebra, and starbuck, create shifting spatial patterns. The animals come together and then disperse as they travel each day from grazing sites to water sources and back. Some drink from small puddles, while others join the elephants at the larger pools. Staying together helps to protect each animal from predators.

Since a predator can eat only one prey animal at a time, the larger the herd the lower the chance that any particular herd member will be eaten; however, the larger the herd the more likely it is to attract predators.[28] We went on a photography safari to the Okavango Delta, and two of our guides gave different accounts of the collective fusion and fission of temporary grazing groups of zebra. One said that zebra are timid, so they tend to hang around with large herds of other ungulates. The other said that zebra are smart and choose to stay around large groups of other prey that are slower than zebra to run from a predator. The two accounts reflect interesting differences in perspective: in the first the zebra are expressing their internal temperaments, while in the second the behavior of the zebra is functionally linked to the collective behavior of the other ungulates and the predators. Either way, the movement of the zebra is related to and influences the movements of other species.

For many collective spatial patterns of animals, we do not know how the individuals interact so as to regulate the pattern. Electric eels in the Amazonian rain forest herd shoals of fish into a ball and then work together to stun, kill, and feast on them.[29] At night in the shallow water of the Pacific near the Baja coast, a group of jumbo squid form a long vertical helix that surrounds a school of fish, compresses the school, and makes it accessible as prey for all of the squid.[30] Both eels and squid are considered solitary animals, but they interact so as to come together and coordinate their movement with the movement of the fish. The success of both eels and squid further depends on how the fish interact spatially with each other to compress their school into a compact shape that can be herded.

The Regulation of Activity

In many natural systems, the collective regulation of activity produces task allocation, which adjusts the distribution of effort into types or tasks.[31] In the development of an embryo, this is differentiation, the process that assigns function to cells by setting which cell becomes which type. As cells jostle around, the rate at which they encounter certain chemicals, both in the medium they move through and on each other's surfaces when they touch, determines the functions that they take on.[32] Some cells, such as stem cells, have many options for what they can become later, while other less labile cells are likely to continue in the same function.[33]

In some animal groups, task allocation coordinates when some individuals watch out for predators while others feed. Ostriches lower their heads to dig for roots and insects to eat (not to hide from unpleasant realities). They sometimes hold their necks up to look for predators such as lions. The larger the group of ostriches foraging together, the less time each one has to spend being vigilant for lions and the more time it can spend feeding with its head in the sand.[34] An ostrich with its head down cannot see how many other ostriches are nearby, so ostriches interact visually when their heads are up. Apparently when an ostrich sees a neighbor, it is likely to keep its head down a little longer on its next feeding bout.

Social insect colonies use task allocation to regulate effort in various colony tasks to adjust to changing conditions.[35] Which worker does which task, and which workers are currently active in performing a task, are regulated by interactions among ants. Task switching has been demonstrated in many ant species, and differences among species in task switching reflect differences in ecological conditions.[36]

My research on harvester ants shows how colonies shift allocation to various tasks in response to the current situation. When I began observing harvester ants, the first step was to classify the behavior of ants outside the nest into four tasks: nest maintenance, midden work, patrolling, and foraging. Nest maintenance workers are mostly at work inside the nest, plastering the inner walls of the chambers with moist soil that dries to an adobelike surface and carrying the dry sand out of the nest. They make short trips outside the nest carrying sand or refuse, put it down, then go back into the nest. Midden workers sort refuse, such as the husks of the seeds that the ants eat, into piles on the nest mound. The nest mound is covered with tiny pebbles brought in by the ants, and the midden workers mark these pebbles with colony odors. Patrollers come out early in the morning, move around the foraging area, and then go back to the nest. The safe return of the patrollers stimulates the first wave of foragers to leave the nest, taking the directions chosen by the patrollers. Foragers travel from the nest together, forming streams or trails of ants. Then each forager diverges from the trail to search for a seed; once it finds one, it returns to the trail to carry the seed back to the nest, where it puts down the seed and, depending on its rate of contact with returning foragers, may go out on another trip.

Ant workers switch from one task to another, in response to feedback from their interactions with other ants and to the conditions associated with a task, such as a food source or an obstruction at the nest. When a harvester ant in a laboratory colony encounters midden workers at a high rate, it is likely to switch to do midden work.[37] When extra food becomes available, a nest maintenance worker, patroller, or midden worker switches tasks to forage. When more patrollers are needed, for example, because of a disturbance or an incursion by foreign ants on the nest mound, nest maintenance workers switch to become patrollers. Although ants change tasks, not all such changes are possible.[38] When more nest maintenance workers are needed, as when there is a mess on the nest mound to clean up, none of the other workers switch back to nest maintenance. Instead, new nest maintenance workers are recruited from the younger ants inside the nest. Thus, there is a one-way flow of workers from the younger nest maintenance workers into foraging, and once an ant is a forager, it stays a forager.

Honeybee workers also switch tasks. A worker performs tasks such as brood care and nest maintenance inside the nest when younger and then begins to forage outside the nest. When a young bee becomes a forager depends on how many older foragers are present. If the number of older foragers declines, young bees switch to foraging sooner. This is regulated by interactions between older and younger bees that influence the hormonal condition of the younger bees. Without sufficient interactions with older bees, the younger bees become foragers as their hormonal conditions shifts.[39]

Interactions regulate not just which task an individual performs but also whether an individual is currently active. As described in chapter 2, a harvester ant forager goes out on many trips a day, and each time it returns to the nest it then decides whether to leave again. This decision is based on its rate of encounters with returning foragers with food.

In brains, the collective regulation of activity by groups of neurons is a form of task allocation. Certain neurons are committed to particular functions. The visual system of vertebrates is a well-known example; particular neurons that respond to visual images are located in much the same place in the brains of different animals. But the function of other neurons can change, and this is determined by interactions among neurons. For example, in fruit fly brains, neurons perform "multiplexing": the same neurons are used in different combinations to accomplish different tasks.[40]

Regions of the brain differ in how tightly function is linked to particular cells. For cognitive tasks, fMRI tests on different people, or on the same people at different times, often show that particular functions are performed by groups of neurons but that the locations change; thus, the neurons engaged in cognitive tasks tend to be scattered across the brain.[41] Differences in the fidelity of neurons to particular tasks depend on function; the activity of particular neurons varies more for cognitive functions than for more ancestral functions such as vision and olfaction.[42] Thus, neurons, like ants, switch tasks or functions, and interactions among groups of neurons regulate their activity.

Movement patterns and task allocation are only two of many diverse forms of collective outcomes in natural systems, all arising from interactions among the participants that respond to changing conditions. How interactions produce collective outcomes can be described by outlining the steps in a process, such as the series of biochemical reactions between interacting molecules that set the rate of a metabolic pathway, or by a mathematical model that specifies how the outcome changes over time. The next chapter discusses models of collective behavior, particularly natural distributed algorithms.

4

Beyond Emergence

How do the interactions of individuals in the aggregate produce collective outcomes that respond to changing conditions? The intriguing problem is packed into the phrase "in the aggregate," which covers, in a vague way, how the actions of the individuals generate the collective outcome.

The term "emergence" has a long history and has been used in many ways, but in general it refers to this problem: the difficulty of explaining some phenomenon or process from the behavior of its components. In the mid-1900s, ideas about emergence were introduced by philosophers of science, such as Carl Hempel, in a sober effort, devoid of any mysticism, to build a theory of explanations. In their view, what is emergent is what cannot be explained by looking at the parts.[1] Thus, emergence is always relative to a particular explanation. It is the tip of the iceberg of the whole, emerging above the water, while the rest of the explanation, the list of the parts and how they work, lies below and out of sight. The idea of emergence has since taken many forms, but it is never offered as a description of how a system works; instead, it always refers to the lack of an explanation.[2]

As a graduate student looking for ways to talk about collective behavior, and more generally about alternatives to reductionism in biology, I used ideas about emergence to refer to those aspects of ant colony behavior that we cannot explain by looking at ants as independent of each other. My evoking of emergence deeply annoyed some of the professors in my dissertation committee, who said that this was nonsense, or at best merely philosophy, not science. Now emergence is trendy, and I have become, like some of the members of my PhD committee, though for different reasons, a curmudgeonly professor who would prefer not to talk about emergence. The concept isn't helping us understand anything better, and we are ready to replace its mystical glow with something more substantial: let's figure out how to explain collective behavior, and then there will be no need to invoke emergence.

Increasingly, "emergent" is used merely to identify as collective the behavior that arises from interactions among participants in relation with the current situation. Several currents came together to generate the growing interest in emergence and collective behavior. One is the explosion of technology based on distributed computer systems that now dominates our lives and has made emergence commonplace. Though it is easy to forget that we depend on the collective behavior of neurons in brains to type an email message, it's obvious that the email's arrival at its destination relies on the collective behavior of a network of electronic interactions that links the computers of the sender and receiver by way of the routers and servers in between.

The revolution in technology that led to the internet brought together many ideas about outcomes that arise from relations and thus cannot be predicted solely from the state of the parts. Statistical mechanics in physics, beginning in the 1920s, showed that the behavior of systems of many components is not fundamentally deterministic. World War II stimulated the growth of cybernetics: the study of systems with feedback in which the action of the system changes the environment, in turn changing the action of the system. Alan Turing showed how, with simple binary choices, a computer can perform sophisticated operations. In the 1970s, the mathematician René Thom explored systems that bifurcate, suddenly transforming from one state to another instead of changing in gradual linear steps, and it became clear from the work of Robert May and others that chaos is a predictable outcome of alternative possible states. Around the same time, physicists began to work with the spin glass, a model of realignments of electrons to explain changes in phase, from gas to liquid or liquid to solid, of the whole system. In the 1980s, artificial intelligence blended with neuroscience; models of parallel distributed processes predicted how a network of neurons could produce a change of state in a brain by using simple local stimulation that merely caused one neuron or another to fire or not like Turing's computer.

Then, in the 1990s, all of these ideas, about dynamic systems that arise from the untraceable contributions of many elements into global outcomes, very quickly burst into everyday reality, as we became inextricably enmeshed in Google, the internet, and cell-phone networks.

Part of what drives current interest in emergence is that, as the technological systems on which we rely expand beyond central control, it seems increasingly important to understand their limits. Emergence has some frightening forms, from relentless streams of army ants to AI algorithms that dominate their creators. As in the animated movie *Big Hero 6*, where particles are harnessed together into a swarm that attacks as a weapon, simple agents in the aggregate can form an adversary that is much more powerful than the agents.

In biology, interest in emergence is fueled by the growing sense that we have reached the limits of reductionist biology, which embodied the hope that we would come to understand how natural systems are regulated simply by specifying which parts interact. The molecular revolution of the twentieth century had enormous success in identifying the parts of many interesting and important systems and the mechanisms that link them, such as the steps in the transcription of genes, the components of the immune system, the pathways that change molecules from one to another and create energy in metabolism, or the dozens of different types of cancer. Without this massive effort to identify the components of each system and the mechanisms that connect them, we could not have arrived at the recognition that all of this knowledge is not enough. That recognition is now leading to a paradigm shift in biology. Besides knowing the means of interaction, we need to know how the interactions are linked with the current situation so as to allow the system to adjust as the situation changes.

Because collective behavior is made up of the actions of its participants, trying to parse behavior into individual and collective components never works. This futile effort, as Rémy Chauvin wrote, is *s'oriente[r] dans une voie sans espoir*, which could be translated as "leading down a blind alley," or as "trying to find one's way on a hopeless path." It divides the system artificially.[3] The part of the iceberg above the water is not different from the part below; the waterline is determined by the position of the viewer. For example, the question of whether an ant colony has a "colony mind" has no good answer. The behavior of an ant colony is the behavior of the ants. Trying to figure out which part of the colony is the ants and which is a colony mind, the tip of the iceberg, is a hopeless quest, because the behavior of the colony is the ants working together.

The framework developed by Denis Walsh, which divides the group and participants into two levels of agency, provides a way to bypass the blind alley.[4] We can say that a flock of birds can turn in response to a predator, and that a bird within the flock turns in response to a change in the direction of its neighbor. For an ant colony, we can say that the colony chooses how much to forage in response to the current availability of food, or that an individual forager decides to leave the nest on its next trip in particular conditions, such as the rate at which other foragers are returning with food. The decisions of the foragers produce the decision of the colony, but the process that the foragers use to make decisions does not refer to some overall decision of the colony. A forager does not decide how much the colony should forage today, but only whether to leave the nest to forage right now.

When we can describe how patterns of encounters among individuals adjust outcomes in changing situations, we have explained collective behavior and it is no longer emergent. Then we can move past emergence to explanations, and abandon the idea that collective behavior is added on top of individual behavior to create something extra.[5]

Collective Behavior as an Interaction Network

To explain how a particular form of collective behavior works is to specify how the participants interact, how these interactions are related to the current situation, and how the interactions produce the outcome. Though many early models of collective behavior sought to reveal common, basic principles that would lead to a single general theory of collective behavior, emergence, or self-organization, it is now clear that the processes that generate collective behavior are very diverse. By now many different models have been developed to describe the myriad processes in nature that use interactions to generate collective outcomes.

It is important to remember that similar collective outcomes are not necessarily the result of the same process. Birds evolved ways of responding to each other's movements that allow a flock to turn in response to predators or obstacles. Cells evolved ways of moving together to generate layers of tissue that link to and avoid others in the course of development. A murmuration of starlings careening in the sky may look like a group of migrating cells, but those two outcomes are produced by different regimes of interactions and responses to surroundings. They evolved from different evolutionary pathways. In the same way, though the eye of a mammal and the eye of a crab look alike and seem to do the same thing, they evolved through very different trajectories and they function differently.

Because there are always many different processes that could have the same outcome, even when a model predicts an outcome that resembles the one we see in nature, it may not describe the process that is actually at work. Moreover, since the same process can have different outcomes in different conditions, a model that fails to predict the outcome could be incomplete rather than wrong.

Despite these caveats, models are very useful empirically, especially when they are slightly wrong so that their predictions do not match observations. The exercise of figuring out why predictions do not match observations can show us what we do not understand and helps to confirm that the model describes the process that is actually generating the observed outcome. This exercise proceeds by iteration: making a model (or mathematical description of what is happening in the system), asking how observations differ from the model's predictions, revising the model, and so on. Richard Levins says in *Evolution in Changing Environments* that a model can be either so detailed that it merely replicates what we have already observed or so general that it does not apply to anything.[6] Modeling is the attempt to stay between those two extremes, so as to find ways to ask empirically how a system works.

The first models of collective behavior were a response to the skeptical and reductionist environment of the 1980s and '90s. The quantitative study of collective behavior began with insights from sociology[7] and physics[8]; early studies focused on social insects[9] and animal movement.[10] The aim was to demonstrate that a system can achieve some predictable outcome, such as a spatial pattern, without central control. That this is possible has now been established beyond question, generating many new journals, conferences, books, and university interdisciplinary centers. Two excellent introductions to this modeling effort are David Sumpter's *Collective Animal Behavior* on models of collective behavior, and Nancy Lynch's *Distributed Algorithms*, which outlines the investigation of distributed algorithms in nature, blending computer science and biology.[11]

Very broadly, models of collective behavior can be classified into two types.[12] One type is a top-down model that predicts the outcome of rates of interaction among participants without necessarily specifying how the response of each individual to interactions generates the outcome. The mathematics for describing and predicting change has developed many branches since its origins in calculus in the eighteenth century. An excellent introduction to dynamical systems in biology is Fred Adler's *Modeling the Dynamics of Life*,[13] and James Ferrell's *Systems Biology of Cell Signalling* introduces the dynamics of behavior inside cells.[14] Spatially explicit models of ecological processes led to the use of drift-diffusion models for animal movement, based on random walks.[15] Another example of this approach is the work of Nicholas Ouellette and his colleagues showing the influence on midge swarms of long-range acoustic signals: each midge makes a sound through the beating of its wings.[16] Moving in response to the sound created by the whole swarm, each midge keeps moving toward the strongest sound intensity, which changes as the midges move.

Control theory is another method for making top-down models of collective behavior for systems with nonlinear dynamics in which interaction rates are linked by feedback.[17] Such models were developed in engineering to predict how change in the rate of one operation could affect another—for example, in a factory where the production of one component influences the production of another. This is discussed in more detail in chapter 5, which describes how feedback in collective behavior is associated with changing conditions.

Modeling interactions as networks leads to top-down metrics that can predict collective behavior, such as the spread of a contagious disease. A network is a set of entities, called nodes or vertices, that are linked by some relation. One such metric is the distribution of the numbers of links or edges per node. Variation among nodes in the number of links produces some subgroups that are more linked to each other than to the rest of the networks. The pattern of links is the modularity of the network, which predicts how information or disease will spread around it. An epidemic in such a network will spread quickly within the subgroups that interact most with each other and more slowly between other subgroups. Chapter 6 discusses how the modularity of interaction networks is associated with changing conditions.

The second type of model is bottom-up. Agent-based models that specify individual interactions and calculate an outcome from those are bottom-up. For example, Iain Couzin and his colleagues have developed models that explain how fish schools move together by specifying how the movement of fish determines how likely they are to interact and how the interactions then change the way they move.[18] Another agent-based model used cell-phone data on how people move to estimate the probability that they would meet, and then how the encounters influenced the probability of infection with Covid-19.[19]

A distributed process is a general form of a bottom-up model of collective behavior. In engineering, distributed processes were developed to accomplish tasks in large systems with many interacting components, such as computer programs and data networks. Distributed algorithms are agent-based models that describe how a collective outcome is achieved through a distributed process.[20] In a set of individuals, such as computers, people, ants, or cells, each individual has a status, or current state. They interact using messages, which can be very simple. The individual reacts to interactions with itself and with others and responds to stimuli from outside the system, which can change its state. A system like this can vote or make choices out of a set of options, find the shortest path through a network, or create an ongoing communication system.

Many forms of collective behavior can be described as a distributed process.[21] For example, bacteria use quorum sensing to regulate their collective behavior according to density.[22] Bacteria can detect chemicals in the matrix where they are swimming around. When each bacterium secretes a chemical cue, its concentration reflects the number of bacteria present. At some threshold concentration of the substance, corresponding to a threshold density of bacteria, all the bacteria tend to change behavior, creating a collective outcome, such as moving away or gathering together to form a mat that clings to the wall of an intestine.

Distributed processes use stochasticity to regulate collective behavior.[23] Noise, or randomness in the components of a system that interact, helps to initiate spindle formation,[24] jostles amino acid polymers around to facilitate protein folding,[25] and introduces obstacles to the spiral waves that allow slime molds to aggregate.[26] Another example is in collective search, when members of a group, such as firefighters, robots, or ants, search for something that could be anywhere in an unmapped region, such as a person in a burning building or a patch of food. Random movement can increase the probability of discovery, because it sends searchers in different directions.

Turtle Ant Trail Networks as a Distributed Process

Turtle ants use a distributed process to maintain and repair trail networks through the vegetation in the canopy of the tropical forest. To describe this we used an agent-based model in which simulated ants moved around a simple array or graph of linked nodes.[27] The model had two parameters that could be adjusted. The first parameter was how quickly the pheromone decays: the more quickly it decays, the higher the rate of flow needed to reinforce each choice at a junction so as to leave enough pheromone to get the next ant to make the same choice. The second parameter was the probability that an ant leaves the trail to explore: the more often ants leave the trail to explore, the more likely they are to find new food sources, but also the more likely they are to get lost.

When we tried different values of the parameters in simulations, we found that the values that worked best were similar to those we had observed in the field. This result does not prove that the ants operate as in our model, but it does mean that the distributed process in our model is consistent with what the ants are actually doing. However, like any model, this one is not complete. The ants travel through a tangle of branches and vines, with each node a unique shape. The physical configuration of the nodes in the vegetation determines how quickly a particular edge is reinforced.[28] In the simulations we used, all nodes are equivalent, so the simulations did not reflect all of the environmental conditions to which the ant trails are responding. The next step in modeling the process the ants are using will be to consider the effects of variation among nodes in how likely ants are to use them.

Nestmate Recognition and Immune Response

Nestmate recognition in ants is another distributed process. It depends on olfactory interactions in which one ant senses the odor of another by smelling the cuticular hydrocarbons on the other ant's body. One explanation for how this could work is that each ant carries a colony odor that acts as a passport. When one ant encounters another, it could check whether the other has the same colony's odor. This raises the question: How would an ant identify another ant's passport? One suggestion is that an ant may check its own odor somehow (though it's not clear how, since its antennae do not reach its own body very easily), compare its odor to the other ant's, decide how different the two odors are, and react aggressively if the other ant's odor is too different from its own. However, this account is not consistent with what we observe: the extent of difference between colonies in hydrocarbon signatures does not predict aggression.[29] A second problem with the passport explanation is that there do not seem to be invariant or colonywide odors: an ant's cuticular hydrocarbons change over time, and not all individuals in a colony have the same cuticular hydrocarbon profile.

Fernando Esponda and I modeled an alternative explanation based on a distributed process.[30] This process allows colonies to form a collective response to other colonies, although no ant recognizes ants of every other colony as different. In our model, an ant assesses the odor of another and locates that odor relative to a boundary that divides the space of possible odors into nestmate and not-nestmate odors. This boundary is arbitrary. It is not based on any judgment by the ant about similarity to its own odor. Most important, we consider, first, that this boundary changes with the ant's experience of encounters with other ants, and second, that the ants differ in their nestmate–not-nestmate boundary. Initially an ant accepts all odors as nestmates. Young ants tend to stay inside the nest, so every other ant that a young ant meets is likely to be a nestmate. Then, as an ant gets older and works outside the nest, it encounters ants from other nearby colonies, and something unpleasant may happen; for example, the ant's neighbor might recognize it as a stranger and attack it. These encounters with ants of other nests tend to shift the ant's boundary so that it does not accept all odors as those of nestmates.

Our model of nestmate recognition as a distributed process, rather than one that relies on a static and inherent ID, helps to explain a puzzling result of many experiments on nestmate recognition. Such experiments usually consist of putting some ants from colony A in a container with ants of colony B. If they fight, we say that that at least one side is recognizing the other as foreign, though we do not know if the recognition is mutual. But if they do not fight, we do not know whether they all are failing to recognize that the other ants are not nestmates or they just are not in the mood for fighting. The results of such experiments are notoriously variable; for example, combinations of ants from the same two colonies sometimes fight but sometimes do not. Interestingly, the outcome is more predictable, and aggression between ants of different colonies more likely, in experiments with a larger number of ants.[31] This trend can be explained if the ants are using the distributed process we describe rather than a static ID or passport. The larger the samples of ants from colony A and colony B, the greater the likelihood that the sample will include some ants for which the odor of the other colony is on the foreign, non-nestmate side of their recognition boundaries.

Shelby Sturgis's experiments on nestmate recognition in harvester ants suggest that ants change their recognition boundaries over time.[32] He put ants from two harvester ant colonies together in a small container to see if they would fight. Task groups differed in their responses to encounters with ants of another colony. Foragers were much more likely to fight than nest maintenance workers, reflecting differences between foragers and nest maintenance workers in their likelihood of meeting ants of another colony. The older foragers meet the ants of neighboring colonies almost every day when their respective trails overlap many meters from the nest.[33] By contrast, the nest maintenance workers, which are the youngest ants to work outside the nest, rarely go beyond the edge of the nest mound; they travel just far enough to put down some refuse on the midden before going directly back in. It seems that over time, as ants change tasks, their opportunities for contact with ants from other nests increase, making them more likely to change the recognition boundaries that they use to distinguish nestmates from others.

The distributed process described here for nestmate recognition in ants is analogous to the process used in the adaptive immune system of many vertebrates. The reaction at any time depends on the particular configuration of immune cells (or ants) that happen to encounter a pathogen (or ant of another colony.) In the immune system, as new pathogens are encountered, new antibodies are formed and T cells with receptors specific to these pathogens are selected. This is how vaccines work; creating encounters between immune system cells and a new pathogen leads to the production of antibodies that will fight that pathogen if it shows up again. At any particular time, each pathogen can be recognized by only a tiny fraction of the immune cells. The lymphatic system helps in the search for pathogens, sending out T cells that find pathogens and call in the response (through a different distributed process that regulates searching by T cells).[34] In the same way, ants of different colonies fight only if the ants that meet are ones that identify the other as a non-nestmate.

Another similarity between ant colonies and the immune system is that the baseline is tolerance, which then shifts to reaction in response to interactions with others. Just as the default for young ants seems to be to accept every ant they meet as a nestmate, so in a developing mammal the default is to accept other cells, such as those of its mother, as belonging to the same body. In fact, as Thomas Pradeu argues in *The Limits of the Self*, the main role of the immune system is not to defend a hard boundary between self and not-self, but to participate in distributed processes inside the body that manage inflammation, wounds, relations with the microbiome (which constitutes many of the cells in a body), and other internal systems.[35] The response to pathogens is part of an ongoing collective conversation among immune cells and many other kinds of cells, of which only some are pathogens from the outside and antagonism is only one of many functions. Similarly, in an ant colony, engaging with ants of other colonies is only one of many tasks, and for the ants working inside the nest, tolerance of each other is the rule.

The Evolutionary Ecology of Collective Behavior

The original goal of early models of collective behavior was to show that emergence is not a mystery; the process that generates collective behavior can be explained. That effort, which led to models of many different systems, was so successful that it is now clear there are many ways to build collective outcomes from interactions. This raises the question of whether there are ecological patterns in the diversity of processes that generate collective behavior. How does evolution shape how collective behavior works?

Collective behavior evolves through the ways that individuals interact. Interactions generate collective behavior that, like any phenotypic trait, has ecological consequences for the participants (discussed in detail in chapter 9). These ecological consequences determine how natural selection shapes the interactions that produce collective behavior.

An example of ecological constraints on collective behavior is found in the shape of bird flocks. Hangjian Ling and his colleagues modeled the interactions that allow jackdaw flocks to change shape, and their work points to physical constraints that affect all birds.[36] They argue that turning by changing wing angle, in response to the movement of neighbors, uses less effort by a bird than adjusting speed, which requires it to flap its wings. If the birds adjust to each other by turning, the flock can change shape as a wide row, with each bird beside its neighbors, while if the birds adjust to each other by speeding up or slowing down, the flock can change shape as a line, with one bird behind another. Ecological factors may lead different bird species to use interactions to coordinate flock shape in different ways, according to conditions that might support either turning or adjusting individual speed. In a species that moves through heterogenous spaces, squeezing between obstacles, a bird may adjust its speed in response to its neighbor's position, to create a line or a narrow shape. By contrast, for a species whose flocks tend to move through open spaces, a bird may just turn in response to its neighbor' position, which requires less energy but creates a wider shape. For all of the birds in a flock, the world that the flock is flying through sets up the consequences of the interactions that they use to stay together.

Another example of collective behavior that clearly has ecological consequences for the group is task allocation in a social insect colony. When a new food source is available, the colony that allocates more foragers may get more food; when the nest is damaged, deploying ants to fix it may save the larvae from harm. Differences among species in the interactions that generate task allocation reflect differences in the changing conditions to which colonies adjust.[37] The magnitude and rate of shifts in worker effort depend both on how often and how much the environment changes and on the costs of incorrect adjustment.[38]

Because a group's capacity to respond to changing conditions matters ecologically, it seems likely that similar dynamics in the environment have supported the convergent evolution of the dynamics of interactions that produce collective behavior. In other words, how individuals use interactions to generate collective outcomes is likely to reflect the changing conditions that the behavior responds to, because selection acts on how a collective outcome adjusts to conditions. The next two chapters outline hypotheses about how collective behavior evolves in relation with its surroundings.

5

Rate and Feedback

In this chapter and the next, I outline hypotheses about correspondences between how collective behavior works and the changing surroundings in which it functions, and describe examples that illustrate these patterns. My goal is to propose new questions about how interactions are used to adjust collective outcomes to changing conditions. These hypotheses are meant as a framework for research on collective behavior, not as a deductive theory; I expect that these trends will not hold in every case, and that these correspondences do not cover all of the ways in which collective behavior is associated with changing conditions.

To characterize how collective behavior is regulated using interactions, I first draw on two basic ideas from dynamical systems: rate and feedback. In the next chapter, I discuss modularity, a concept from network theory. In this chapter, I suggest how rate and feedback are likely to reflect the conditions to which collective behavior responds. The rate of adjustment is how quickly interactions change in response to conditions. The feedback regime sets the way that interactions stimulate and inhibit activity.

To characterize how environments change, I use ideas familiar in ecology: stability, the distribution of resources or demands on the system, and energy flow. Stability broadly describes both how often and how much the environment changes. The distribution is the pattern in time or space of the resources that the system seeks, or of the system's needs or demands that it must meet. To simplify this discussion,

I consider a continuum of distributions of resources or demands, from scattered and random to clustered, occurring in patches in space or bursts in time. The stability of the environment is associated with the distribution of resources or demands. Clustering or bursts in time correspond to unpredictable or infrequent events, while resources that appear at random, though not homogenous, are equally likely to appear everywhere and thus may be associated with higher stability than patchy resources.

Energy flow, which describes how the environment determines what is needed to carry out a collective process, always depends on many dynamic processes, but again, in order to consider broad trends I reduce it to a simple gradient. At one end is the situation in which what is expended is high relative to what is obtained, so the ratio of energy out to energy in is high; these are conditions in which the behavior uses a lot of resources. At the other end is a situation in which what is expended is less than what is obtained, so the ratio of energy out to energy in is low; in this situation it is easy to engage in the behavior.

This chapter outlines the first two hypotheses, introduced in chapter 2, that suggest how the rate at which interactions are adjusted, and the feedback regime that links interaction rates to activity, both reflect the stability, distribution of resources and demands, and energy flow of a changing environment.

The Rate of Interaction

How quickly rates of interaction are adjusted is perhaps the most basic and obvious correspondence between the dynamics of behavior and the dynamics of environment. Change in this rate is the fundamental means of regulation in nature.[1] In living systems, hardly anything happens only once, and the rate at which it repeats matters. The heart sends a wave of blood down the aorta at many beats per minute, and a drastic change in heart rate has widespread effects because that rate affects interactions with many physiological processes. Living systems are always ticking.

Rates of interaction can change in response to conditions. Such shifts are usually nonlinear. A linear change is incremental: the same amount of stimulus always gets the same response. Linear dynamics are rare in nature, except in the short term. For example, more gene expression means more protein synthesis, but only for a while; eventually protein synthesis levels off. Most collective outcomes involve feedback that creates nonlinear responses to interactions. A response is nonlinear when the amount of response varies with the amount of input or of the condition it depends on. Anyone who has spent a day with a two-year-old has seen how emotional response increases nonlinearly with fatigue: the child is easily distracted from a thwarted desire when not tired, while the same setback leads to a meltdown when a nap is needed.

Collective outcomes often depend on combined rates of interaction. For example, as inflammation increases, the rate of production of cytokines is increased, changing the rate at which immune cells are summoned. Another example is the regulation of the foraging activity of honeybees, which use the rate at which returning foragers meet a nectar storer to unload their nectar by regurgitating it. This rate is linked in turn to the amount of stored nectar, which is itself the result of a process of accumulation.[2] When there is more stored nectar, the bees that receive the nectar take longer to get to the hive entrance to meet the returning foragers. Then the rate at which foragers leave the nest is related to their experience while waiting to unload: the longer a forager has to wait, the longer it stays inside the nest, after unloading, before it goes out on its next foraging trip.

Changing rates of interaction drive changes in the population numbers of different species that live together as an ecological community. "Interactions" in the ecological sense are those that lead to changes in population size, such as when one species eats another. A well-known example of the outcome of combined rates of interaction in an ecological community is the oscillation in the numbers of Canada lynx and hares, their prey. (This is an example of the effect of combined rates on population numbers in an ecological predator-prey interaction, not of the collective behavior of lynx and hares as they chase, evade capture, and get caught.) The oscillations in population numbers over generations of lynx and hares arise from the relations among the rate of reproduction of both species and the rate of predation. When lynx are plentiful, they eat hares at a high enough rate to drive down the hare population, with the result that eventually lynx run out of food for long enough that their numbers go down. When the number of lynx decreases, the hares are not getting eaten as much, so they can reproduce quickly enough to increase their population numbers rapidly. This increase then provides enough prey for the lynx population to grow large again, enabling lynx to eat enough hares to eventually bring the hare population down again. Thus, the numbers of lynx and hares continue to oscillate.

Another example of the effect of changing rates of interaction in an ecological community is the impact of fishing on coral growth. Fishing influences the collective behavior of coral reef fish, which eat the algae that compete with coral.[3] Coral reefs compete for space with algae, and algae become established when coral dies. Many coral reef fish eat algae, making room for the coral to grow. Coral reef fish move together in groups and tend to follow each other into new feeding areas. To get into a reef to feed, fish must meet each other at a threshold rate and join together as small feeding groups. When large fishing boats remove massive numbers of fish, the remaining fish experience a rapid and large decline in the rate at which they meet other fish. Then many reefs have no fish herbivores because the population of fish is not dense enough for feeding groups to form and make their way together into the reef. The absence of fish leads to higher algal growth, which crowds out the coral. Loss by fishing of the same number of fish but at a lower rate can allow the fish to continue feeding on the algae on the reefs, when there are enough fish to maintain a rate of contact that generates the collective behavior that leads them into reefs.

The rates of interaction among cells and molecules, as well as among organisms, produce collective outcomes. How cellular processes differ in rate is likely to be associated with their environmental conditions. For example, cellular processes that work by diffusion and those that work by trigger waves differ strongly in how they adjust their rates. Diffusion is the result of chemical interactions that push a substance from higher to lower concentration, such as when ink is spilled into the middle of a puddle and spreads to the edges. Diffusion is slow, and it becomes slower and weaker the farther it gets from the source because the concentration of the diffusing substance becomes more similar to that of the solution it is in. Trigger waves, by contrast, are rapid and steady. They work through linked feedback loops that propagate in a wave as molecules trigger reactions in neighboring ones. This interaction maintains speed as the wave moves; the wave is just as rapid far from the source of the spreading substance as it is near it.

Trigger waves that generate a rapid cellular process are used in environmental conditions where a rapid response is important. For example, trigger waves induce cell death in unfertilized frog eggs, speeding up the development of fertilized eggs.[4] Tadpoles grow in ponds among masses of fertilized eggs. The speed of development of fertilized eggs is probably important for tadpole growth. Competition with other tadpoles inhibits tadpole growth.[5] This competition may favor processes that quickly cull eggs that are not viable, allowing some eggs to grow large sooner than the rest of the cohort. Trigger waves also lead to rapid regeneration of fish scales, which are needed quickly because of the danger of infection when there is a gap in scale cover.[6]

In general, collective responses are likely to adjust quickly when the surroundings change quickly or often, or when there is a high cost to not responding quickly. The rate of formation and dissolution of mixed-species flocks of birds sets the rate at which they find new resources. Birds that search together for rapidly changing resources are quick to form temporary groups. How quickly they join together determines how quickly they can learn from each other about the location of new food sources and how quickly they can arrive there.[7]

The effect of rapid adjustment of an interaction rate is familiar in human collective behavior. In Paris during the French Revolution, the dissemination of speeches by reading the texts at local Jacobin clubs allowed action against the aristocracy to amplify rapidly during the Reign of Terror.[8] Before that, when news took many weeks to spread by diffusion from one person to another, political change occurred more slowly. In the twenty-first century, the high speed of communication by internet allowed for the rapid organization of protests during the Arab Spring of 2010, and also in response to the 2020 murder of George Floyd by police in Minneapolis.

The rate at which interactions are adjusted is likely to depend on the stability or frequency of change in the surroundings that are relevant for that behavior. The stability of the environment is associated with risk, which combines two different aspects. The first is the probability that something bad will happen; for example, the risk of getting stuck in traffic when driving at rush hour is much higher than the risk of being abducted by aliens. The second is the cost of the adverse event if it does happen; in most cases being stuck in traffic is merely annoying, while being abducted by aliens would have far-reaching and drastic consequences. This second aspect of risk, its cost, is associated with the feedback regime: when the risk of bad consequences is high if the activity is stopped, it may be worthwhile to set the default at continuing the activity.

Philip Grime's work on plant physiological ecology captures this distinction by defining stress and disturbance according to the two components of risk: the likelihood of some kind of harm and the magnitude of that harm.[9] Stress is ongoing but not drastic; by contrast, disturbance is intermittent and damaging. Grime defines "stress" as the suite of conditions in which extra work is required to regulate some necessary process. For example, when water is scarce, plants have to work harder to manage water loss. One way they do this is to open and close their stomates, the pores that allow water to go in but that also lose water to evaporation. The extent to which stomate size is regulated in different species is associated with the sustained risk of running out of water. Another response to risk is shown by pea plants, which grow more roots when nutrient availability is variable, thus increasing the chances of capturing water if needed.[10] "Disturbance" goes beyond stress to outright damage; for plants, it is the loss of tissue to herbivores, trampling, or other events that kill some or all of the plant. Some plant responses to this level of risk are broad dispersal and rapid reproduction.

How individuals deal with risk has been studied extensively in economics and in the decisions of individual animals.[11] Collective behavior also reflects the riskiness of the surroundings. In gene transcription networks, long cascades can be used when the threat of rupture is low enough to allow sufficient time for the many interactions needed to adjust transcription; in riskier conditions, shorter cascades are needed.[12] In some animal groups, such as squirrels or meerkats, an individual that sees a predator may make an alarm call that warns others. Alarm calls are risky for the caller, who could attract the attention of the predator, but might save many others. The collective behavior that regulates this trade-off has evolved in response to the probability and magnitude of the risk: how often predators show up, how likely they are to eat a group member, and how much the alarm call helps the listeners avoid predation.

The evolution of metabolic rates in dogs shows an association with the stability of food supply and the risk of starvation. Dogs and their wolf ancestors differ among lineages in metabolic response, an interaction of cells and their products that is a form of physiological collective behavior.[13] Huskies evolved from an ancestral wolf species that could metabolize starches slowly, so they can feed on carbohydrates, which provide a sustained source of calories when food availability is low. Because huskies are able to deal with instability in the food supply, they can pull sleds and travel for long periods. By contrast, the ancestors of many other dog species are wolves of a different species that lived on meat. The metabolism of these dogs evolved in conditions of stable food availability with less risk of starvation. They need more frequent meals, and their metabolism does not slow down in conditions of scarcity, so they cannot survive when feeding on the slower timescale that can sustain huskies.

The rate at which interactions are adjusted to the current situation is also associated with the distribution of resources or demands. When resources or demands are clustered or arrive in bursts, in time or space, rapid mobilization may be required. By contrast, a stable system allows the luxury of inflexibility. The rate at which interactions are adjusted is related to risk. Speed in obtaining a resource is important if there is a risk that the resource will be lost to competition. Danger can be clustered in time or space, such as when a predator appears suddenly, and collective responses to such danger are likely to rely on rapid interactions. However, when resources are steady or scattered in time or space, or when risks are low, there is no need for a quick response, and persistence might be more important than speed.

To summarize, the first hypothesis, illustrated in figure 5.1, is that rate is related to stability and the distribution of resources or demands. The rate of interaction corresponds to the rate of change in the surroundings to which the behavior is responding. The more stable the conditions, the more slowly interactions are adjusted in response to the current situation; the more unstable the conditions, the more rapidly interactions are adjusted. The rate at which interactions respond to the environment also depends on the distribution of resources or demands: the more clustered they are in time or space, the more rapid the adjustment to allow for immediate response.

The Rate of Interaction and the Stability of the Surroundings

This section provides further examples of slow and rapid adjustment in interaction rate, corresponding to the stability of surroundings and the distribution of resources and demands in space and time.

Slow Adjustment

Spatial memory in rodents. Slow changes in conditions are associated with the slow adjustment of neural connections to form memories. The collective behavior of interacting neurons generates new memories. Slow interactions among neurons form new circuits to consolidate memory during sleep, on the timescale of twenty-four-hour cycles.[14]

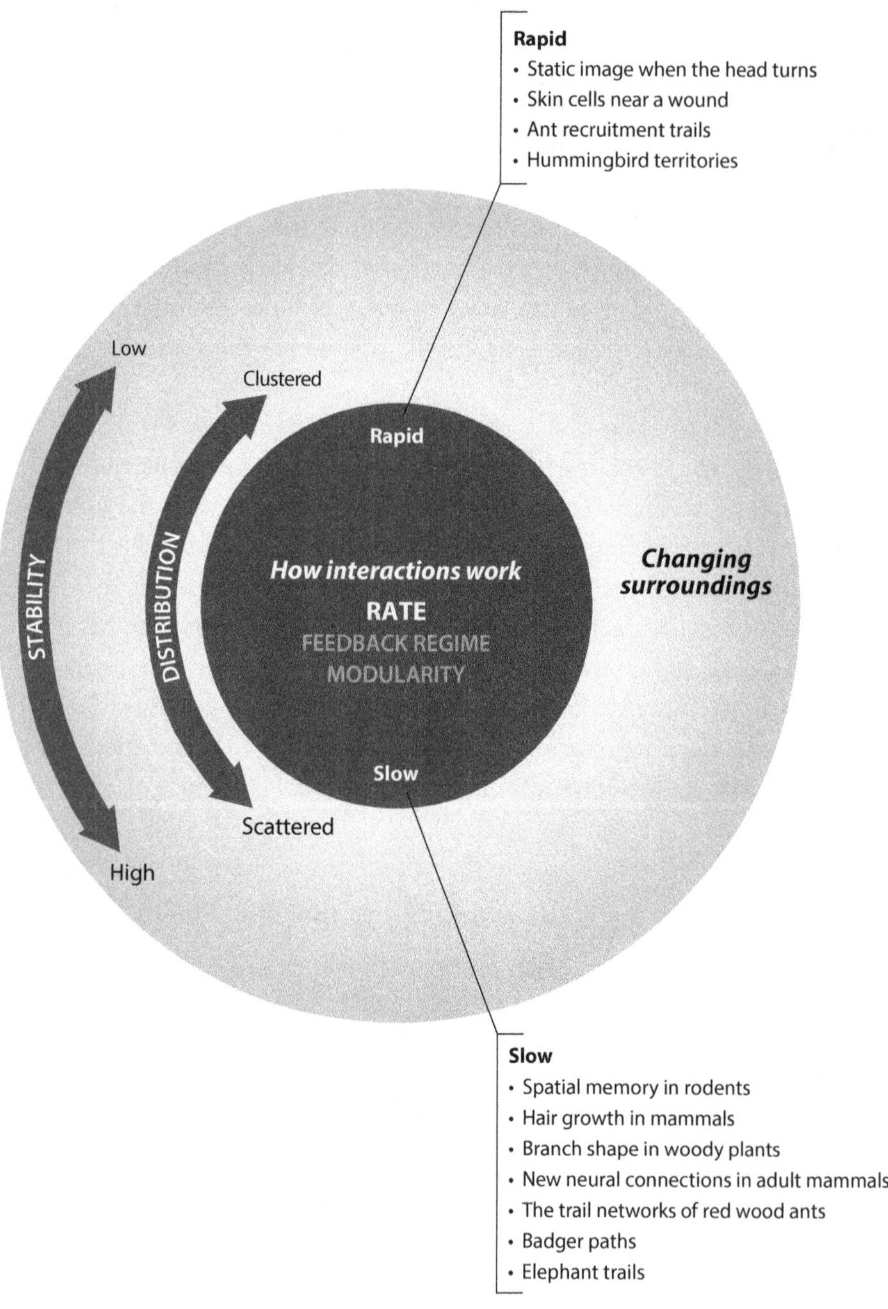

FIGURE 5.1. Rate of interaction and changing conditions. The symbols are the same as in figure 2.1 (see page 11). The figure illustrates hypothesis 1, that the rate at which interactions are adjusted is associated with the stability and distribution of resources and demands in the surrounding conditions. Slow rates are associated with stable conditions and scattered or random resources and demands; rapid rates are associated with unstable conditions and resources and demands that are clustered in time or space. Examples in the text that illustrate this, taken from chapter 2 and from the following section, are listed on the right.

These slow interactions correspond to spatial features of the rodents' world that change very slowly, on the scale of days and months, such as vegetation and barriers to movement. Memories based on a twenty-four-hour sleep cycle would not track a more rapidly changing spatial environment.

Hair growth in mammals. Slow changes in interactions among cells lead to slow response in hair growth. Interactions among hair and epithelial cells lead to the growth of new hair. Hair cells grow slowly, on a timescale of weeks that is based on slow periodic cycles of two signals: one that inhibits growth and one that promotes regeneration.[15] Hair growth in mammals evolved as a response to seasonal changes on the scale of months, to have more fur during the cold season. Fur is a response to a gradual change in temperature that happens very slowly, so the cellular interactions that produce hair do not have the capacity to speed up to grow quickly.

Branch shape in woody plants. Slow shifts in cell interactions allow woody plants to make changes in form. Interactions among tissues in plant stems generate the form of woody plants as they grow, determining how branches set their angles and how they change shape in response to obstacles or stress from wind. In flowering bushes and trees, the wood that changes shape, called "reaction wood," uses tension that pulls the branch upward. Shifts in position stimulate the growth in reaction wood of fibers with gelatinous walls that contract and pull nearby fibers along with them. These slow changes in shape as the branch grows arise from the capacity to sense a very stable feature of the environment: the position of the growing branch relative to gravity.[16]

New neural connections in adult mammals. Neural connections change slowly in a stable environment. In general, sensory neural function in adult mammals relies on well-established sets of connections among neurons that grow over a long time and respond slowly to change.[17] The slow development of neural connections may be related to the stable surroundings of a brain. In most vertebrate animals, brains sit inside hard cases; traumatic injury is rare, and brains are not well adapted to recover from changes in the network of connections among neurons.

The trail networks of red wood ants. Some ant species forage for stable resources that are consistently available, and they change foraging trails slowly. Red wood ant colonies live in nests that can persist for many decades, housing successive generations of colonies, in the pine forests of northern Europe. The ants feed on the sugary excretions of aphids that are established in the pine trees, where the aphids do not move around very much. The interactions that set the direction of red wood ant foraging trails can adjust only very slowly. Colonies rarely form new trails, and even a new food source does not stimulate the ants to form a new trail once the food is gone.[18] The interactions among the ants that maintain the trail networks produce only slow changes in the number of foragers allocated to each trail that leads up to the aphids in a particular tree. Foragers are drawn onto a trail by an encounter with an older forager that used the trail the previous year.[19] This yearly timescale of adjustment through interactions between older and younger foragers corresponds to even slower changes in the location of the trees where the ants find aphids.

Badger paths. European badgers change paths very infrequently, on a timescale much slower than that of human development. Groups of European badgers live together in setts that they build together, with underground chambers and tunnels.[20] A group uses the same sett over many generations, and paths from the sett to foraging areas are used for decades. Because this collective process does not rapidly adjust the site where the group lives or change the paths to and from the sett, the badgers do not modify their paths quickly enough to respond to rapid changes in the landscape created by human development. In England, tunnels for badgers have been built under the highways to allow new generations of badgers in a particular sett to continue to use the same pathways without crossing the highway and getting killed by cars. Changes in the spatial organization of the badgers' sett and paths are adjusted in response to the slowly changing landscape of a forest, scaled to the long life cycles of trees.

Elephant paths. Similarly, elephants move in large groups of long-lived individuals along well-established routes that persist for decades.

This behavior evolved in a stable environment in which the location of water holes and vegetation rarely changed. The low resilience of their trail system is tragically mismatched with the rapid changes in the availability of resources brought about by clearing and farming by people. One effort in Kenya to move elephants to a reserve failed because the elephants could not change to a very different route; when taken to a reserve about two hundred kilometers away, some of the elephants walked back to their original route.[21]

Rapid Adjustment

Static image when the head turns. Rapid adjustment of interactions among neurons allows for rapid adjustment of the perception of an image when the head turns. The visual system detects shifts when the head turns, but interactions among neurons compensate for this so that the image appears to stay still. The firing of neurons adjusts extremely rapidly to head movement. Some neurons send a particular signal that negates the signals from other neurons that would generate the perception of a moving image. These compensating neurons work with astonishing speed, at a timescale of about 120 milliseconds, to counteract the ones that show the image to be moving, thus creating a static image.[22] This adjustment by interactions among neurons must be as rapid as the actual change in the view of the surroundings when the head turns.

Skin cells near a wound. Interactions among epithelial skin cells close a wound rapidly.[23] Triggered by a variety of signals, the skin cells of a fruit fly near the wound margin adjust their current interactions by beginning to elongate and move toward the wound. (Some of the changes are not only in rate; for example, some cells fuse together, creating cells with many nuclei.) An important adjustment is an increase in the rate of contact as cells gather to plug the gap created by the wound. Wounds are clustered, rather than distributed homogenously in time, because they occur sporadically and are also risky: failure to respond quickly enough can incur the high cost of leaving a body open to pathogens and further injury. Like first responders to a fire, the cells adjust their interactions rapidly to deal with events that occur quickly and unpredictably.

Ant recruitment trails. Rapid change in olfactory interactions allows rapid response to new, short-lived food sources. Some ant species can rapidly form trails to a picnic in the time it takes you to eat half of a cookie and provide a windfall of crumbs. Such species are opportunistic, specializing on ephemeral resources, such as a picnic or the nectar of flowers, that occur in patches and could appear anywhere at any time and that many ants are required to retrieve. To form a recruitment trail, they must increase interactions, in the form of deposited trail pheromone, at a rate that is set by the rate at which the pheromone decays. Opportunistic recruitment requires a rapid change in interaction rate so as to mobilize others to the food source.

Hummingbird territories. Rapid adjustment of territorial interactions among hummingbirds leads to rapid shifts in their use of patchy and ephemeral food sources. Frequent and rapid interactions lead them to shift locations and to decide whether to fight with each other.[24] Hummingbirds burn calories at a very high rate and must keep eating to survive. They forage on food sources that are clustered in both space and time, such as groups of flowers with nectar that may be available for only a short time.

These examples illustrate the hypothesis that the rate of interaction is likely to adjust more quickly in unstable conditions or when resources are clustered in space and time. Feedback, a means of adjusting rates, is considered in the next section.

Examples of Feedback Regime in Collective Behavior

The second hypothesis is that feedback regimes are associated with the stability of conditions and the energy flow that the environment requires. The *Merriam-Webster* dictionary defines "feedback" as "the return to the input of a part of the output of a machine, system, or process." Feedback is most familiar in processes with input and output, such as electrical flow or a chemical reaction. Here, however, I consider more broadly any process that depends on interactions, such as the movement of a group of cells or birds, to ask in what conditions feedback is likely to inhibit or to stimulate activity.

Feedback is widely used in natural systems to regulate changes in rate. For example, the contact guidance of cells depends on positive feedback from interactions among moving cells and fibroblasts.[25] In the course of development, cells move in streams from one location to another. Elongated fibroblasts act like rails along which the streams of cells move. When a cell sticks on one side to a fibroblast, a signal is stimulated that makes the cell more likely to form a protrusion along that side. This feedback, from the rate of contact to a change of shape and of movement, encourages the cell to move forward and continue to adhere to the fibroblast guide, and the process continues as the cell moves along.

A system can combine different forms of feedback that work together to regulate it. A suite or network of feedback loops is a "feedback regime" consisting of the linked processes that stimulate and inhibit activity, such as a signaling pathway in a cell or a set of relationships in a social group.

Here I contrast the conditions associated with two broad classes of feedback regimes: those with the default set for activity to stop unless stimulated (figures 5.2A and B), which I will call "default not to go," and those with the default set for activity to continue unless inhibited, which I will call "default go" (figures 5.2C and D). The second hypothesis is that this aspect of a feedback regime—whether the default is to go or not to go—is likely to be associated with the energy flow required to deal with a situation. When it takes a lot of energy to run the system relative to what it brings in, making the ratio of energy out to energy in high, it is expensive to keep the system going if it's not needed. Then feedback can prevent activity if it's not necessary. Positive feedback saves costs when the system is inactive unless something pushes it to go; the default is to not engage in the process until stimulated. But when activity is cheap, making the ratio of energy out to energy in low, the activity can continue unless there is some reason to stop it. The default is to keep going unless something negative happens.

The warning and signal lights on a car dashboard illustrate this relation between energy flow and the default state set by the feedback regime.[26] One kind of warning is associated with positive feedback; its function is to instigate some drastic and costly response that is out of the ordinary—for example, by telling you that the oil is low or the brakes aren't working. This warning calls for action: from the default state of not even thinking about the oil or the brakes, you immediately have to deal with the situation. The other kind of warning is associated with negative feedback: you don't have to do anything, and the default state is to stay the same, but there is something that you might want to turn off. An example is the dashboard light showing you that the headlights are on. No urgent action is required; the function of the dashboard light is merely to remind you that you have the option of turning the headlights off.

A familiar biological example of this distinction is the difference in what it takes to gain or lose fat. In the production of fat, the default is to keep going, converting excess energy into lipids to be stored. This default evolved in an environment where starvation was a persistent threat and it was worthwhile to make fat whenever extra resources were available, so as to have it ready to use in times of scarcity. By contrast, the process that metabolizes fat must be activated by the movement of muscles; the default is not to break down fat unless resources are needed.

Positive feedback is associated with systems in which the default is not to go, because it stimulates or increases activity. The two figures in the top row of figure 5.2A illustrate a process with the default not to go. Figure 5.2A represents a process that is not active unless it is triggered by some cue. Figure 5.2B adds positive feedback, showing a process that, once stimulated to go, leads to interactions that stimulate further activity. If someone looking up at the sky on a city street stimulates others to look up, creating positive feedback, the number of people looking up will increase. Positive feedback leads to amplification, a nonlinear response, when it causes each further interaction to stimulate a stronger response. Ants at a nest respond to the amount of pheromone they detect; as the number of ants on the trail increases, so does the amount of pheromone deposited, and that, in turn, increases the rate at which ants leave the nest.[27] Recruitment rates can slow down when there are no more ants waiting in the nest to join the trail.

Negative feedback is associated with a process with the default set to go, because it inhibits or decreases activity. When each interaction magnifies the decrease in the rate of activity, the rate will decline exponentially, a nonlinear change. For example, when an infectious disease is denied opportunities to spread, negative feedback is created and the infection will decline in the population.[28] Because each infected person can cause many other infections, negative feedback through reduced contact can lead to an exponential decline in the rate at which people are infected.

Figures 5.2C and 5.2D illustrate processes that continue unless they are inhibited, either by a cue or by negative feedback. The default is to go. For example, turtle ant foragers keep walking along their trail through the vegetation but stop in their tracks if they encounter an ant of the genus *Pseudomyrmex*, an occasional predator, and they stay frozen for about a minute. When ants stop moving, they also stop putting down trail pheromone, and the previously deposited pheromone evaporates, so the rate of flow on the turtle ant trail decreases. If the behavior of the *Pseudomyrmex* is not influenced by the behavior of the turtle ants, then their presence provides a cue that slows foraging but there is no feedback, as shown in figure 5.2C. However, it is possible that the *Pseudomyrmex* ants can detect and follow the pheromone trails of the turtle ants. If so, then the higher the rate of flow the more likely that the predator will be attracted to the trail, thus slowing the trail as the turtle ants stop moving. This is an example of the negative feedback shown in figure 5.2D.

In physiological processes, regulation arises from some combination of feedback that generates inhibition or excitation. A signaling or metabolic pathway links interactions, some stimulating and some inhibiting, among different molecules. John Tyson and his colleagues

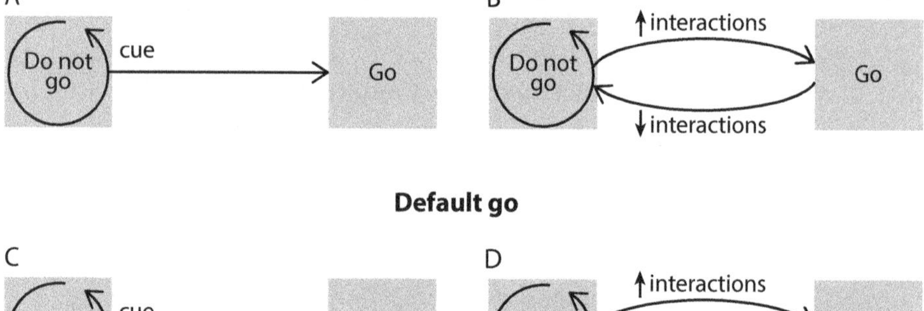

FIGURE 5.2. Feedback regimes. The top figures illustrate a feedback regime in which the default is not to go and to remain inactive unless stimulated to go. A illustrates a cue that stimulates activity. B illustrates positive feedback that stimulates activity; interactions stimulate the system to go and a decrease in interactions leads back to inactivity. The bottom row shows a feedback regime in which the default is to go. C illustrates a cue that inhibits activity. D illustrates negative feedback that inhibits activity; interactions lead the system toward inactivity and a decrease in interactions leads back to activity.

describe the feedback regimes of molecular regulatory networks as "sniffers, buzzers, toggles and blinkers" that turn processes off and on or lead them to oscillate.[29]

Homeostasis occurs when a system continually adjusts back to a baseline or set point. This is a form of feedback regime. An idealized example is a sine wave, which fluctuates around a mean or baseline. A process regulated as a sine wave is adjusted down toward the baseline at the peak of the wave, and adjusted up toward the baseline at the trough of the wave. Homeostasis is associated with an equilibrium or optimal stable state that a system seeks to maintain.

Although homeostasis is the most familiar form of feedback regime, it is only one of many that regulate the outcome of collective processes. The outcome of a feedback regime might not be to keep a system the same but instead to make the state of the system different from what it was before, pushing it back and forth between one state and another, or irreversibly toward a new one.

The resilience of a system is set by the feedback regime, which determines what it takes to activate or inhibit the activity, and whether and how it returns to a set point if disturbed. As the system moves around or changes, the amplitude of its change reflects what is at stake in keeping it at the set point. Boats provide a mechanical example.[30] A boat's resilience in response to the turbulence of the water it is in combines both primary and secondary stability. A rowboat has high primary stability because it stays flat on the water, but it has no secondary stability: in high waves it cannot adjust. A kayak, by contrast, has high secondary stability: it is able to lean sharply from side to side and still recover. On the open ocean with strong waves, a kayak is safer than a rowboat, but a rowboat works well on a quiet lake.

A biological example of resilience as a consequence of feedback is hemoglobin production, which is regulated in response to elevation, because more red blood cells are needed at high elevation when there is less oxygen in the air. Hemoglobin, whose titers differ in resilience in populations at different altitudes, is regulated in relation with how much and how frequently people change elevation, since the risk of insufficient oxygen is high if the hemoglobin concentration is too low. In people living in the Andes, who travel across a great variety of elevations, hemoglobin is adjusted across a broader range than it is in people who live at sea level and do not change elevation.[31]

A common form of feedback regime that stimulates activity is an excitable system, which can be generated by combined positive and negative feedback.[32] Each interaction has some stimulating effect, increasing the likelihood of activity, but the impact of each interaction decays over time, in some cases because of negative feedback from the activity. The response accumulates over many interactions (figure 5.3): if enough interactions occur before the last one decays, the accumulated response reaches the threshold for an action to happen. But if the next interaction occurs after the last one has decayed, then the process must begin again with the next interaction.

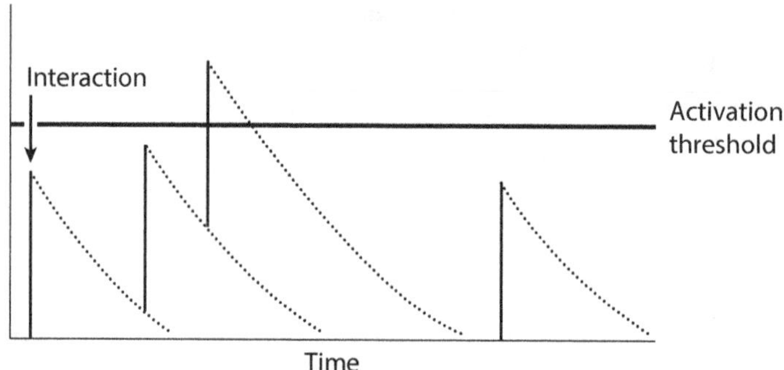

FIGURE 5.3. Excitable dynamics. Each interaction contributes to a response. The effect of each interaction decays over time. The effects of successive interactions accumulate over time, so that a response occurs when interactions occur often enough to reach a threshold that is likely to stimulate activation.

In an excitable system, the rate of decay sets how quickly the process responds, and thus how committed it is to the default state of inactivity. Neurons are the canonical example of excitable dynamics. A neuron decides whether to fire by accumulating stimuli from other neurons through axons and dendrites, the long projections that connect them. The decay of each stimulus is caused by the leaking of the electrical charge from the axon. Once enough stimulation has accumulated, the neuron fires. How quickly a neuron accumulates enough stimulation to fire depends in part on the rate of leakage; the more rapid the leakage, the slower the accumulation.

Just as neurons use the accumulation of stimuli from other neurons to decide whether to fire, harvester ants use the accumulation of stimulation from antennal contacts to decide whether to leave the nest to forage.[33] Their behavior reflects a neurophysiological process whose mechanism is still not known; this process is stimulated by an antennal contact and decays over time, on the timescale of seconds. The rate of decay sets how quickly an outgoing forager leaves the nest in response to the rate at which it experiences antennal contacts. At the peak of the daily foraging period in a large colony, with returning foragers coming into the nest at a rate of about two to four ants per second, it takes an outgoing forager about thirty seconds to accumulate enough antennal contacts to decide to leave the nest. If the effect of each interaction decayed more slowly, the effect of interactions would accumulate more quickly, and the colony could increase foraging activity more rapidly. The more rapid the decay, the more interactions required to stimulate activity.

Ant pheromone trails provide another example of an excitable system that can be tuned to conditions. Trail pheromone decays through evaporation. Ants follow each other when enough pheromone accumulates; once the pheromone has evaporated, no ants follow and the trail ends. The rate of pheromone decay determines how committed the ants are to a default state of inactivity, because the decay rate sets how likely ants are to follow a new trail. If the pheromone evaporates very rapidly, a high rate of flow is needed to create a new trail. By contrast, pheromone that evaporates slowly promotes rapid formation of new trails. This suggests that the more ephemeral the resources, the higher the decay rate is likely to be, since more persistent pheromone could lead ants to a food source after it is gone.

To consider how the feedback that regulates collective behavior is associated with changing environments, I collapse the astonishing variety in nature of forms of linked feedback into two very broad categories of feedback regime: those that set the default to go and those that set the default not to go. The second hypothesis, illustrated in figure 5.5, is that the feedback regime is related to the stability of conditions and to energy flow. In stable conditions that change slowly, or where energy use is high relative to the energy brought in, the system waits for the opportunity to act. Positive feedback, in which interactions from some activity trigger further activity, can be used when the default is set at inactivity; action is stopped unless something positive happens, thus saving energy until action is worthwhile. By contrast, in unstable conditions that change frequently and require that the system respond quickly, or when the activity gathers more energy than it uses, the system keeps going unless something inhibits it. Negative feedback, in which interactions from some activity

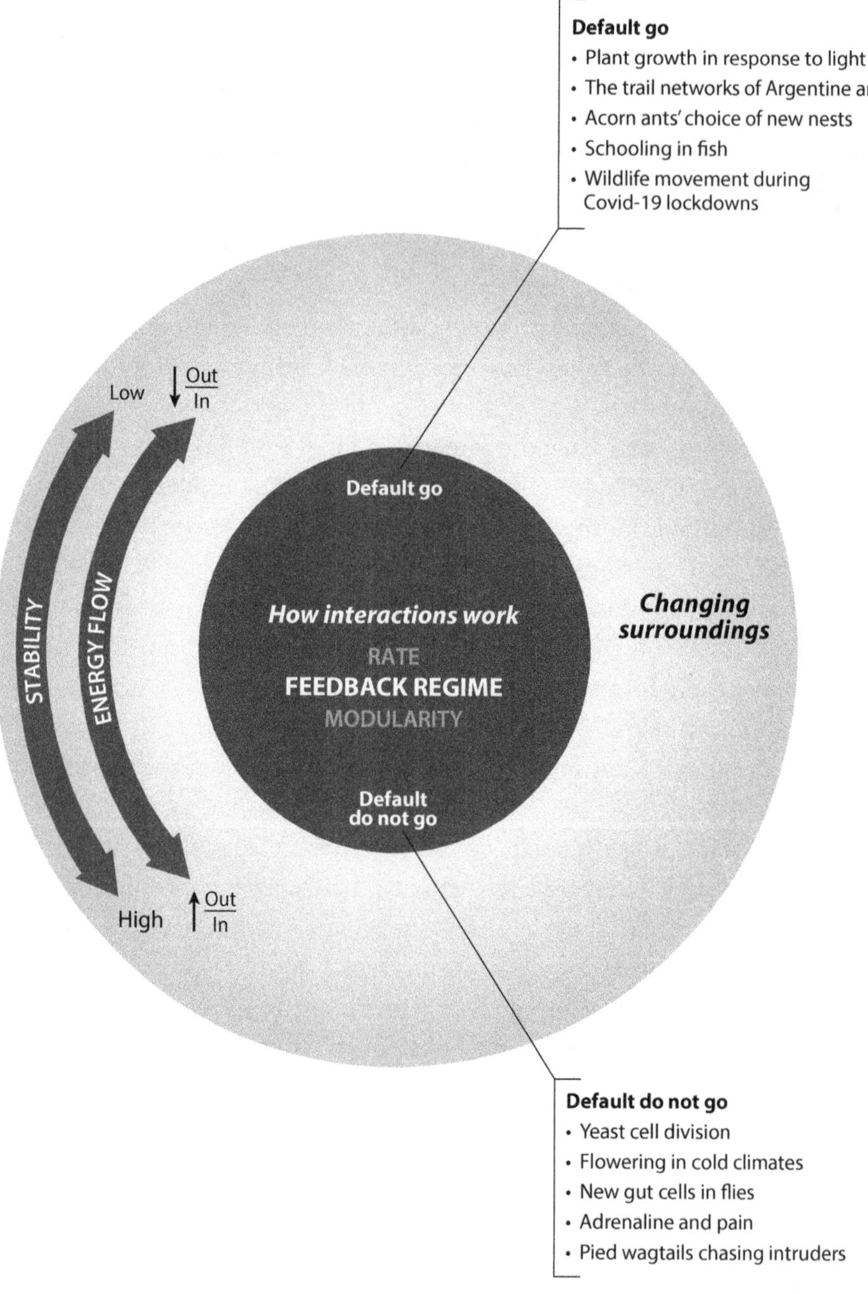

FIGURE 5.4. Feedback regime and changing conditions. The symbols are the same as in figure 2.1 (see page 11). The figure illustrates hypothesis 2, that the feedback regime that links rates of interaction, and activates or inhibits activity, is associated with the stability and energy flow required in surrounding conditions. In stable conditions that change slowly, or where energy use is high relative to the energy brought in, the system waits for the opportunity to act and the default is set at inactivity (not to go). By contrast, in unstable conditions that change frequently, where the system has to make rapid responses, or when the activity gathers more energy than it uses, the system keeps going unless something inhibits it; the default is set to go. Illustrative examples in the text, taken from chapter 2 and from the following section, are listed on the right.

inhibit further activity, can speed up adjustment to rapidly changing conditions when the default is set to keep going unless something negative stops it.

Examples of Feedback Regimes: Default Do Not Go versus Default Go

This section provides examples of how the default for activity, determined by the feedback regime, is associated with stability and energy flow.

Default Do Not Go

A collective process in conditions where the ratio of energy expended to energy obtained is high, or there is a high risk of adverse events, is likely to use a feedback regime with the default set to remain inactive until stimulated. This saves the system from spending resources or risking danger when the situation does not warrant it.

Yeast cell division. For a yeast cell, the default is not to divide, and division is triggered by cell growth. The interactions of biochemical signals within and outside yeast cells regulate cell division and the formation of other cells. The work of Andreas Doncic and Jan Skotheim shows that, in yeast, the decision to reproduce by dividing is based on a switch that uses a positive feedback loop, with the default set at not dividing.[34] Growth induces cyclin-dependent kinase activity, which drives the synthesis of further cyclin proteins to beget even more cyclin-dependent kinase activity. The cell does not divide until it reaches a threshold level of cyclin-dependent kinase. The size of a cell is important because it sets the amount of surface, which collects resources, including oxygen, relative to the volume in which the resource molecules must be distributed. It seems that the feedback regime functions to keep the cell from dividing until it is so large that its offspring cells will be large enough to capture enough oxygen. This conjecture is supported by other work examining fossils of different one-celled organisms, *Foraminifera,* which showed that the size of the *Foraminifera* cells has been able to increase over macroevolutionary time as the amount of oxygen in the atmosphere has increased.[35]

Flowering in cold climates. The default is not to flower, and flowering is triggered by cold. A plant's decision to flower is a crucial step toward reproduction. In many flowering plants, flowering is prevented by suppression of transcription at the FLC locus.[36] In some species, cold disables this suppression, leading to flowering. Species vary in the frequency of their flowering, from many times in a year for some perennial species to once a year for annual ones. These species differences arise from adjustments in how long a period of cold is needed to move away from the default and lift the suppression of flowering. When many bouts of cold are needed, suppression is lifted slowly and the plant flowers only after winter has ended. In winter-flowering annuals, only a short period of cold is needed. This system, with the default not to flower, reflects the risk of flowering in the wrong season. If the winter is long, it keeps the plant from flowering too soon; if the winter is short, it allows the plant to flower as soon as it is safe.

New gut cells in flies. The default is for stem cells to remain stem cells, and their transformation to gut cells is triggered by the death of gut cells. The maintenance of the fly gut depends on interactions of stem cells and gut cells. The work of Lucy O'Brien and her colleagues shows that the production of new cells in the digestive system is tightly coupled to cell loss. The death of a gut cell creates feedback that stimulates a stem cell to produce a new gut cell.[37] This process depends on interactions involving the loss of E-cadherin, which gut cells degrade as they die and are sloughed away. The decrease in E-cadherin sends out extra growth factors or mitogens that drive stem cells to divide. This feedback regime sets the default for stem cells to be quiescent unless gut cells die. It seems that keeping stem cells in reserve, available for other uses, is more important than having extra gut cells, and that conditions are generally stable with respect to the need for more cells.

Adrenaline and pain. The default for response to pain is to tolerate it up to a high threshold. The response to pain of the parasympathetic nervous system of mammals is based on positive feedback linking adrenaline, the nervous system, and movement. Danger stimulates adrenaline, which lowers the threshold for nerves to register pain. This feedback amplifies the pain, making a response more likely, while the adrenaline aids fight or flight. These are both costly actions that are not stimulated except in response to unusual and thus possibly dangerous events.

Pied wagtails chasing intruders. In the territorial behavior of these birds, the default is not to chase away intruders, and chasing is triggered by low food supply. Davies and Houston's classic paper on the territorial behavior of pied wagtails shows that encounters among birds create feedback that reflects changes in food supply.[38] The birds eat insects that get caught in small streams and wash up on the banks. Each bird feeds in a small loop around the stream, first going along the bank on one side and then hopping over to go back down the other side. The flow of food is linked to the amount of time a bird devotes to chasing other birds away rather than eating. Sometimes two birds move along the same loop, and sometimes a third shows up. Davies and Houston call the first the "owner," because often the same bird goes back to the same place day after day, and the second the "satellite," because often the owner does not chase it away. The satellite sometimes chases away the third, who is called the "intruder" because it almost always gets chased away.

The amount of food available is set by the rate at which insects are caught in the water and float down the stream. The more food that is available, the less likely any bird is to bother chasing others away. Davies and Houston's article outlines an elegant model of an individual owner's decision to chase an intruder, by calculating when there is enough food for the owner to tolerate the presence of others. The data fit the model's prediction well enough for the authors to conclude that the birds are behaving in a way that maximizes the food for each bird.

It seems to me that the feedback regime for interactions among individuals, chasing or not, makes sense collectively as well as individually. In the aggregate, it generates a collective outcome that is related to a changing food supply. The default is not to chase but instead for all of the birds to hop along the river eating. But when they are hungrier, they are more inclined to chase; the chased birds are then likely to be chased again wherever they go, because on a day with poor food supply, insects are scarce everywhere. In this way a decrease in food supply triggers positive feedback that amplifies chasing and all the birds spend time chasing or being chased rather than eating. A default set to not chasing leads collectively to more time spent eating unless the situation is so dire that chasing is worthwhile.

Default Go

When activity is easy, with low energy expenditure relative to gain, and risk is low, a process can continue unless it is stopped. Here the feedback regime sets the default to keep going, and negative feedback is needed to inhibit or stop the process.

Plant growth in response to light. The default in young plants is to grow toward light, and growth is stopped by too much light. The negative feedback created by light is generated by interactions among organelles within cells, including the chloroplasts and the nucleus. When a young plant experiences too much light, its chloroplasts put out a chemical that changes gene expression inside the cell nucleus. The changes in gene expression prevent the cotyledons, the first leaves from the seed embryo, from expanding and also prevent the stem from elongating toward the light source. The outcome of the negative feedback is to stop growth so that the plant gets less light.[39]

The trail networks of Argentine ants. The default for Argentine ants is to keep following the trail. A colony forms a trail network, a highway system that connects many nests with temporary trails to food sources. The network is developed anew each spring, expands in the summer, and contracts in the fall, coalescing to a single site each winter.[40] The default is to keep the highway system going; ants move continuously along the permanent trails, and some ants leave the trails to search for new food sources. The interactions that form the trails are based on a trail pheromone that ants put down as they walk, like the turtle ant trails described in chapter 2. Drastic negative feedback, such as the loss of enough ants that eventually the trail dries up, is required to stop the trail. For example, Argentine ants are attracted to some odor in the lining of the top freezer compartment of refrigerators. People find large numbers of dead Argentine ants inside their freezer, because ants continue to follow the first ants to find their way into the freezer and die there; the death of some ants in the freezer is not sufficiently strong negative feedback to stop the trail. Only after large numbers of ants have vanished in the freezer does the trail weaken and disappear. Argentine ants are now invasive in Mediterranean climates worldwide. In their native range, where there is frequent flooding, having the default set to continue following a trail takes the colony to safety because, if some ants make it up a tree, the ants behind them probably will too. This behavior did not evolve to deal with freezers.

Acorn ants' choice of new nests. Species of acorn ants differ in the extent to which the default is to move nests, and when the quality of a potential new nest is low enough to inhibit moving. Colonies of acorn ants in the genus *Temnothorax* nest in acorns, which can easily be broken or rot, leading to exposure to predation and harsh weather. Scouts search for a new site, and when they return, the rate of interaction with returning scouts stimulates the other ants to move the colony to the new nest site. Species differ in how choosy they are about new nest sites, showing a gradient in feedback regime that corresponds to different environmental conditions. Whether the feedback is set to a default of go, regardless of how poor the new site is, differs among species. This seems to depend on the energy, in brood production, that is lost in a damaged or unusable nest site, relative to the energy needed to move. Species that live in places with scarce nest sites are more likely to have the default set to move. In these species, it takes stronger negative feedback to prevent a move; as a result, the colony is likely to move into whatever nest site is available, even if it is less than optimal.[41]

Schooling in fish. The default in schooling fish is to stay near other fish, unless some large obstacle breaks up the school. Staying together is effective against predators, which could strike at any time but can eat only one fish at a time. The more fish there are in the school, the less likely it is for each fish that it will be the one eaten. This feedback regime, with the default set to continue aggregating, is not always effective. Whales are huge predators that have evolved more recently than schooling in fish. Some whales loom up on the fish so quickly that the fish do not have time to disperse before the whale consumes the whole school in one gulp.[42]

Wildlife movement during Covid-19 lockdowns. The default for many kinds of wild animals may be to move into human developments, and it seems that this is inhibited only by human activity and traffic. During the lockdowns of the Covid-19 pandemic, the interactions of humans and animals in cities shifted: there were many more sightings of mammals in cities, such as coyotes and deer, as well as an increase in the sightings of birds. That the animals were more active in cities after only a few weeks of shutdown suggests that their default is to adjust their home range to keep spreading, so that they quickly moved into cities when there was no traffic and human activity providing negative feedback.[43] More generally, tracking studies of many species of mammals show that the more densely populated an area is by humans, the less the animals in that area move around.[44] The spatial dynamics of animal groups that tend to expand their ranges, requiring strong negative feedback to limit their movement, indicate that this collective behavior evolved when new space was easily available, long before the human footprint became as massive as it is today.

An Ecological Perspective on Feedback Regimes

To specify how interactions regulate any collective behavior, we need to know how the feedback that links interactions and changing surroundings is set up so as to stimulate and inhibit activity. For many biochemical and cellular processes, we can specify the feedback regime but do not know how it functions in relation with changing surroundings. For example, genes are stimulated to be expressed, making new proteins, by interactions among a series of chemical signals that stimulate or inhibit steps in the process. This process is initiated by gene transcription networks that regulate how segments of DNA are copied into RNA. One common gene transcription network motif is a fast-forward loop (FFL): signal X stimulates Y, which stimulates Z, and Z is not produced unless both X and Y are present. This motif uses positive feedback. FFL motifs thus set the default as not expressing the gene. I would hypothesize that FFL motifs, based on positive feedback, are most common in transcription networks when the operating cost of producing Z is especially high, or when too much Z expression is risky for the phenotype.[45] Negative feedback can be created by incoherent loops, in which one signal activates transcription but another activates a weaker repressor of transcription. This decreases production by repressing transcription once threshold amounts of the product are reached. The system thus bears the cost of producing two products, one of which acts to inhibit the other. Low operating costs may make such network motifs most common for the proteins that are cheaper to produce.

Our understanding of collective animal behavior is mostly the obverse of that for cell and molecular systems: there are many forms of collective animal behavior for which we know the ecological conditions that the system responds to, but not how this is regulated by feedback from interactions. For example, many vertebrate animal groups vary group size in response to changes in the surroundings, such as a change in food availability. When more food is available, red foxes tend to live in larger groups of several females with one male.[46] A comparison of urban, suburban, and rural areas in England by David MacDonald and his colleagues showed that foxes hunt and live in larger groups in places where there is more food.[47] Formation of the group is slow, on the timescale of months. Group size is regulated so as to change greatly in magnitude, though slowly, as it adjusts to significant long-term changes in the amount of available food. Little is known about the interactions that allow this collective outcome, group size, to adjust to food availability. How do encounters among foxes reflect hunger and the abundance of the current food supply, from earthworms to voles? What is the default? To group together but split up when food is low, or to be apart but stay closer when food supply is high?

Studies of mixed-species flocks of birds in a forest in Oxford, England, by Lucy Aplin, Damien Farine, and their colleagues raise other intriguing questions about the feedback that regulates collective behavior. Birds follow each other to food sources, and they learn how to retrieve food by watching each other.[48] Interestingly, the default is to stick to the same food source. Many birds are likely to follow others to the first food source they learn about, even if other food sources are subsequently discovered. This default suggests that food sources tend to be stable long enough to feed the entire flock. It seems that interactions among birds facilitate first copying and then remaining on the first new resource, so that at times a bird is likely to follow another bird to a new source, but later it will not follow another to a different new discovery. What changes the default from sticking to the same food source as before, to following an innovative bird to a new one? Is there feedback based on the quality of the food source, or on how long since that food source first showed up? Or is there some other feature that makes yesterday's find so much more compelling than today's?

In summary, identifying the feedback acting in a collective process suggests how the surroundings constrain it. In the other direction, learning how the process uses and obtains energy suggests what feedback regime the system may use. Positive feedback to ramp up activity may be needed when it is expensive to perform that activity. Negative feedback may be useful when it is important to keep going whenever possible.

Natural systems often involve many sets of related feedback loops generated by interactions among different participants. Together they produce a network of interactions. The next chapter considers how the structure or organization of this network is associated with changing conditions.

6

Modularity

Collective behavior adjusts to changing conditions using interactions among participants. The interactions are the links in a network, which could use chemical, tactile, visual, or some other means, and the nodes are the individual participants, such as molecules, cells, ants, or zebra.

Networks differ in overall pattern, or structure, depending on how the nodes are connected. For example, all of the people I sent email to or received it from today, and all of the people they sent email to or received it from, are nodes in a network whose links are email messages. If I was the only correspondent of all the people I sent email to, the network would look like a star with me in the center; if each person I sent email to corresponded with all of, and only, the same people I corresponded with, the network would look like a regular array; while if some of my correspondents tend to email each other while others do not, the network would have a set of distinctly connected subgroups.

Modularity is one aspect of network structure. Broadly speaking, it characterizes the uniformity of the pattern of connections. Many specific metrics of network structure are related to modularity, such as closeness centrality, which measures the shortest paths from one node to the others, or degree centrality, which measures the number of other nodes to which a node is linked.[1]

A network has low modularity when all the nodes are equally connected to others. For example, in a network of dominance relations among the males of a chimpanzee troop, all the beta males are subordinate to an alpha male.[2] In this arrangement, there is low modularity because the beta males are all connected to the alpha male in the same way.

A network is highly modular when it has connected subgroups of nodes with few connections to other subgroups. For example, the tree of life, or phylogenetic ancestry, is the network defined by the relation of being descended from. For many groups of organisms, this network is very modular; the phylogenetic tree is shaped like a bush, because many species each diverged from a particular ancestor but are not closely related to other branches.

Figure 6.1 illustrates two networks that differ in modularity. In the top image, the nodes are equally connected to each other, so modularity is low. In the bottom image, some nodes are more connected to each other than to the rest of the network, so modularity is high. Since modularity describes the overall distribution of connections, a highly modular network, such as the mammalian system of nerves, can still have a trunk or backbone to which all nodes are eventually connected.

The third hypothesis about the association between collective behavior and its changing surroundings relates the modularity of interaction networks to gradients in stability and the distribution of resources or demands. High modularity is associated with variable or unstable situations or clustered resources or demands. Networks with high modularity can work faster and have fewer links than networks with low modularity, where all participants are equally connected. Low modularity can be used in stable conditions where it can lead to a more thorough spread of information or resources.

The effects of modularity on collective function have been considered in many different fields. Herbert Simon points out the importance and ubiquity of modularity in both human social interactions and biological systems.[3] In developmental biology, "modularity" refers both to a body plan and to the structure of the gene networks associated with a particular trait that influence the plasticity of that trait's development and evolution.[4] Other fields use other names for the same idea. In systems biology, network motifs are groups of connected signal transducers within a larger modular network. In

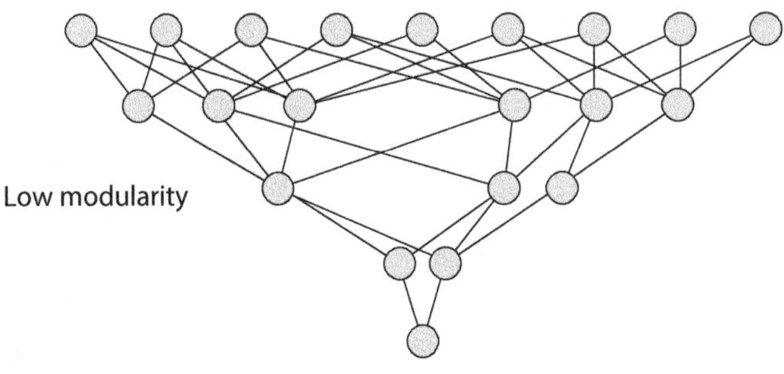

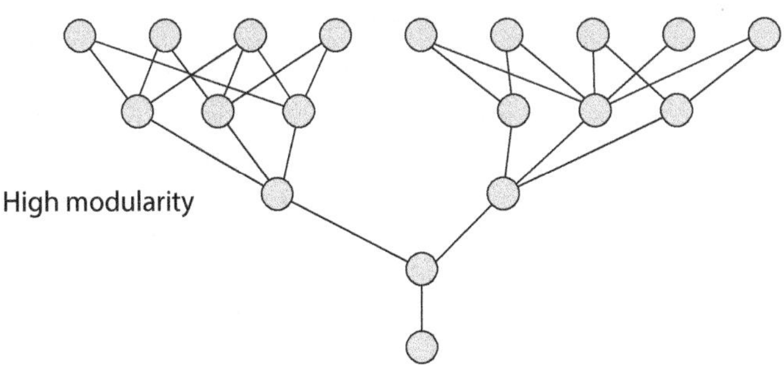

FIGURE 6.1. Modularity. The upper figure shows a network with low modularity: all nodes tend to be equally connected. The lower figure shows a network with high modularity: some nodes are more connected to each other than they are to other nodes.

community ecology, as well as in the growing literature on collective intelligence, modularity is described in terms of the heterogeneity of the nodes in number of connections and in the intensity or function of the links.[5] John Gerhart and Marc Kirschner use "linkage" to describe the modularity of physiological systems, or the extent to which inputs from different signals are linked.[6] In animal behavior, a rapidly growing subfield examines the patterns in the association of known individuals and characterizes modularity in terms of groups or cliques within social networks.[7]

The modularity of a system influences its function in many ways and so reflects the conditions in which it operates. First, modularity influences the resilience of a system. In a centralized system with low modularity, if the central control fails, it can bring down the whole system, while in a highly modular system local failure does not affect other parts of the system. In *The Sciences of the Artificial,* Simon uses the example of two watchmakers, one who constructs the watch in modular subunits and another who assembles it by adding parts to a single core or central unit.[8] If either watchmaker is interrupted while putting together a watch, the one using modular construction is likely to recover more quickly than the other one, because only that module will be affected.

While a task can be accomplished within a module of highly connected nodes without any requirement that the rest of the network be dealt with, high modularity can provide robustness to failure for the whole system. But when the components are interchangeable, not functioning in particular modules, any component can replace any other. If one unit cannot finish a task, another can take it over with an equal probability of success.

Thus, in unstable conditions, when the risk of adverse events or failure is high, more modularity may allow some part of the system to fail while preserving the rest. By contrast, when conditions are stable, or the threat of overall failure is low, or resources are concentrated in one place, low modularity may provide more consistent or widespread control. The response in the United States in 2020 to the beginning of the Covid-19 pandemic provides an example of the advantages of low modularity by showing the disadvantages of its absence. An extreme version of low modularity is centralization, where all nodes are connected to one central node. Because the federal government did not immediately provide a centralized response to the pandemic, the disease spread while more local regulation developed as governors stepped in to respond state by state. The lack of central decisions about the regulation of behavior led to local outbreaks, which then seeded others. The network of opinions about the pandemic and safety was more modular than the system of control, because people within a region tended to share beliefs about the dangers of the disease. Meanwhile, since the actual transmission of the virus was not influenced by belief, high modularity of belief failed to contain the disease, which spread across regions.

The modularity of a system is also associated with the stability of conditions, because the extent of modularity influences the speed of regulation throughout the network. A more modular system can adjust to local changes more quickly than one with central control, partly because high modularity can provide an opportunity to work in parallel, which can go faster than operating in series. In computer software, a quicker outcome is achieved with different modules or cores allowing different units to do the same thing at the same time than would be possible if one unit had to wait for another to finish before proceeding. An example from a natural system is the lungs of mammals, which are divided into lobes, each connected as a separate module to the bronchus on each side. The bronchioles in each lobe are more connected to each other than to those in a different lobe. Each lobe can take in oxygen and host the exchange of oxygen and carbon dioxide. Thus, respiration can operate in parallel in more than one lobe or module. If all exchange came from the same single tube, it would have to travel through the entire lung, working in series.

Another reason that the use of local information within a module can be faster than the use of spatially centralized information is that it takes time and effort to get back and forth from a central location. If a process is completed within a module, the participants elsewhere do not have to share information to arrive at the collective outcome. For example, high modularity can lead to rapid local spread in a disease network. Because our social networks are highly modular, in an epidemic infection spreads rapidly among people in close contact, rather than slowly emanating from one infected person directly to everyone else.

Modularity influences the speed and cost of setting up an interaction network. The higher the modularity the fewer connections overall, since nodes within a module are connected only to each other; the lower the modularity, the more connections because more nodes are connected to each other. Minimizing connection costs is associated with high modularity in development.[9] Simulated networks evolve to be more modular if the cost of making and maintaining connections is high.[10]

Because high modularity can promote innovation, it can allow new responses to unstable or frequently changing conditions. For example,

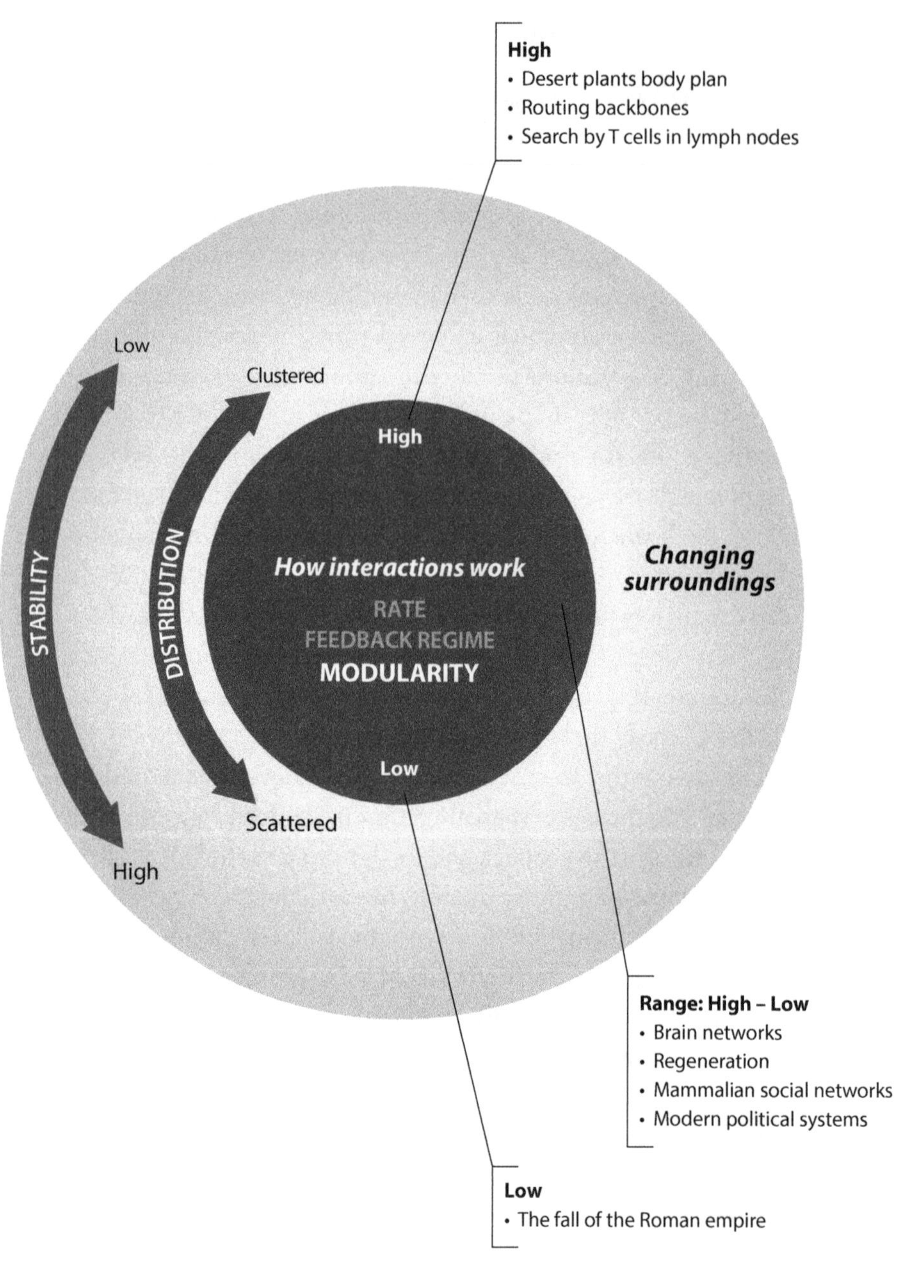

FIGURE 6.2. Modularity and changing conditions. The symbols are the same as in figure 2.1 (see page 11). The figure illustrates hypothesis 3, that the modularity of the network of interactions that regulates collective behavior is associated with the stability and distribution of resources and demands in the surroundings. High modularity, in which subgroups of participants interact more, is associated with unstable conditions, and resources or demands that are clustered in time or space. Low modularity, in which all participants tend to be equally connected, is associated with stable conditions, and resources and demands that are scattered or random in time or space. Examples in the text that illustrate this, taken from the following section, are listed on the right.

learning in human organizations is facilitated when small groups are able to try new methods independently.[11] Another example of innovation through high modularity is the development of body form in plants and animals, which is the result of modular networks of interactions among the cells of the embryo.[12] In developmental networks, traits are often linked through pleiotropy; one set of genes is associated with more than one trait.[13] Over evolutionary history these shared modules of linked genes have led to shared, homologous traits that appear in many species. Evolutionary change in the body form of animals has tended to occur within modules, such as in pairs of limbs, presumably because it is more likely that small changes within a module will be viable than changes that transform many different developmental trajectories with widespread effects on the phenotype.[14]

To summarize, modularity influences both the resilience and the speed of response to changing situations. It is related to the stability of the environment and the risk of rupture, and to the distribution of resources and demands. These correspondences between changing surroundings and the modularity of interaction networks are illustrated in figure 6.2. High modularity is associated with a rapid response to variable or unstable situations, where resources and demands are clustered in space and time. When the threat of rupture is high, high modularity creates redundancy, which can promote resilience. Low modularity is associated with centralized and widespread responses and is more likely in stable conditions or where a more thorough spread of information or resources is needed.

Examples of Modularity in Collective Behavior

This section provides some examples of the association between modularity and changing surroundings. The first few examples are for high or low modularity; the others are examples of a range in modularity that corresponds to a range in the dynamics of conditions.

The body plan of desert plants. Interactions among tissues in plant development produce highly modular networks of physiological processes. As Roberto Saguaro-Gomez points out, high modularity allows for redundancy and promotes resilience in response to stress and disturbance.[15] In some desert plants, one part of the plant can die while the rest survives because resources are moved almost completely within that part of the plant, which has few connections to the rest. By contrast, many plants in more benign conditions regulate the flow of nutrients using a centralized system, so to survive, the entire plant must obtain enough resources.

Routing backbones. High modularity can facilitate rapid communication. In communication networks, it can be more efficient to extend branches from a "routing backbone" to the final destination than to extend directly to each destination from a central source. For example, Argentine ants make networks of trails that link nests and food sources. We found that when they find a new food source off the trail, they recruit ants from the nearby trail rather than returning to the nest to recruit among all ants at the nest. Thus, the interactions related to recruitment occur among only the ants in a small region of the trail near the food source.[16] Recruitment within the local module of nearby ants is faster than going back to the nest. Speed is useful for retrieving resources that are clustered in space and time and could pop up anywhere, anytime, and might be sought by other species. In the same way, the network of nerves in a mammal, which responds quickly to rapidly changing stimuli, is highly modular. Traveling from a node on the spine, a nerve branches out in one part of the body; nerves do not form a centralized network in which each nerve would have to reach its local destination directly from the brain.

Search by T cells in lymph nodes. The modular organization of the lymphatic system in mammals speeds the recruitment of T cells, which search for the antigens on dendritic cells. Interactions of T cells with other cells help to coordinate the search.[17] When T cells are in lymph nodes, each T cell moves within a local network of fibroblastic reticular cells, which provide resources such as the cytokines that promote T cell survival. This network is highly modular; a T cell in an area of the network with many local connections is likely to search that area more thoroughly. The high modularity of the lymphatic system makes immune response more rapid; to trigger an immune response, not every T cell has to tell every other about the pathogens it has recognized. Infection alters the structure of the network of fibroblastic reticular cells, but the network recovers its modular structure after infection, enabling T cells to return to local search.

Brain networks. Brain function is associated with the modularity of networks of neurons. When neurons are connected, one can stimulate another to fire through electrical and chemical interactions. Within a brain, different neural networks differ in the extent of their modularity. There is a trade-off between distance and connectedness. A neuron that interacts with a greater number of other neurons can participate in more and different functions. The shorter the distance between connected neurons, the less energy it takes for one to activate the other. However, connections to faraway neurons require long trips for a signal and provide more opportunities for the signal to be lost—for example, in leakage as the signal travels along an axon. A solution to this trade-off is groups or modules of nodes, each itself a network of neurons, connected to each other.

Function in brain networks depends on the modularity and strength of interactions. Activity within a module is faster than across a widespread network because there is no need for the signal to travel long distances, and strong connections work faster than weaker, diffuse ones. Danielle Bassett and her colleagues showed that perception of visual cues and the response to new events that capture attention arise from strong, localized connections, which easily facilitate change in the activity of nearby nodes. By contrast, learning and cognition are associated with low modularity in the form of weak, long-distance connections. One outcome is that we see faster than we think.[18]

Regeneration. Local regeneration is promoted by modular development. The range of possibilities for regeneration of the organs within a mammalian body is related to stability and risk; if damage is frequent and the cost is high, then regeneration is worthwhile. The capacity for regeneration is a form of high modularity, because it relies on local interactions within an organ among cells that have particular tasks or functions. In mammals, the liver has more capacity for regeneration than other organs. Pericentral cells are a type of liver cell (not a generic stem cell) that can proliferate if needed in response to a signal, Wnt, from other cells.[19] The liver processes toxins that come into the body from outside, and perhaps the modularity of liver cell function evolved to deal with such assaults. Other organs, such as the heart, that are more protected from damage than the liver and the immune system, have lower modularity and no regenerating cells.

Like tissues, species of organisms differ enormously in their capacity for regeneration. Tree species differ in the process that controls growth from the top if the crown is damaged; some trees can grow only from the top, while others can grow locally from anywhere in the tree.[20] *Planarium* is a small multicellular organism with an amazing capacity to regenerate; if cut into 200 pieces, it will grow 200 planaria.[21] Its capacity is not limitless: if cut into more than about 250 pieces, each piece is smaller than the minimum module of interactions among cells needed to stimulate regeneration. It is likely that species differences in the capacity to regenerate are associated with the frequency and risk of local damage.

Mammalian social networks. Mammalian species show a wide range of modularity of their interaction networks. The study of social networks in vertebrates began with Robert Hinde's focus on the role of dyadic interactions in building the structure of primate groups.[22] Dyadic interactions form the links in a broader social network. A familiar form of social network is the dominance hierarchy generated by power relations; an extreme example is a group in which one individual is dominant over all the others. Such a dominance hierarchy would have low modularity since all individuals have the same connection to the dominant one. However, dominance hierarchies in animals often do not conform to this extreme of low modularity because they are not transitive.[23] If A winning a fight with B and B winning a fight with C does not mean that A will win a fight with C, then there is an interesting social network but it is not a hierarchy. More generally, dominance hierarchies are not the rule in mammalian social networks.[24]

Another network with low modularity is not a hierarchy but rather one with many similar links throughout the group, what Mark Granovetter called "weak connections" in humans.[25] Gorillas form stable social networks with low modularity based on widespread connections. Recently, in the San Francisco zoo, a new female was introduced to a gorilla group with the hope that she would mate with the older silverback male. The zookeepers did not begin by introducing her directly to the silverback because he did not exert central control in the group; instead, the sequence of introductions relied on connections with many other individuals. First, the new female was introduced to some of the other females associated with the male silverback, so that she would be accepted by them. After that, she was introduced to the more socially dominant female, and then finally to the silverback. Connections with the females associated with the silverback helped to stabilize the new female's relation with the silverback.[26]

The structure of interaction networks in primate groups is associated with patterns of change in food supply.[27] Comparing the primate groups of different species suggests that the more rapidly the resources change, the more modular the social network becomes, thus allowing for rapid movement to new locations. At one extreme are baboons, which live in large, stable groups of low modularity in the savanna, where resources are stable and scattered.[28] At the other extreme are species that live in tropical forests and forage for fruit, which appears in short-lived patches and thus is clustered in space and time. Spider monkeys, who eat fruit, have a highly modular system with very local fission and fusion of small groups and no allegiance to any individuals that could assert central control.[29] These frequent changes in group composition make it possible to track the rapidly changing, clustered food supply. Similarly, capuchins in tropical forests move around together in small fluid groups, changing their trajectories in response to fruit ripening on trees and the locations of other groups.[30] Forming subgroups or modules allows them to respond quickly when a new patch of fruit becomes available.

The fall of the Roman Empire. Modularity in human collective behavior has been studied from many perspectives. Edward Luttwak argues in *The Grand Strategy of the Roman Empire,* that centralization or low modularity in military strategy brought down the empire.[31] Initially the system was highly modular, as local leaders were bribed or rewarded to act on behalf of Rome to control the locals. Later the empire attempted to create a centralized system, sending soldiers and supplies out to the provinces from Rome. Once the empire was very large, it was too difficult and expensive to respond to challenges far away by bringing people from the distant center. Low modularity was too slow to resist local challenges. The people in the Rhine were able to defeat the Roman army because they could move around locally and attack more quickly than the Roman army could respond.

Modern political systems. Stephen Haber and colleagues show that, over the course of the past three centuries, the modularity of human institutions in different parts of the world has developed in relation with the stability of the environmental conditions that influence food production, storage, and transport.[32]

This work contrasts the ecological conditions related to food supply in the tropics, the steppe, and the temperate zone. In the tropics, food spoils rapidly because high temperatures facilitate the growth of the bacteria that generate decomposition. Where it is not possible to store food, there is no need for long-term contracts related to ownership or for networks of long-distance travel to transport food. Food was traded and consumed in the location and at the time that it was produced. On the steppe, it is too dry to grow crops, and the main source of food was livestock animals, which can consume vegetation and become a source of stored food. There people formed loose groupings of nomadic bands or tribes. In the temperate zone, by contrast, cultures developed that were based on the consumption of pulses and grains, because the growing season is long and food can be stored. However, stability is low because productivity can be limited by catastrophes such as weather extremes, including severe cold, storms, and flooding.

These differences in the stability of food storage led to important differences in the modularity of the political systems that regulate the exchange of information and trade. Many societies in the tropics consisted of small groups, bands, or tribes. The exchange of information was limited because trade took place over small distances, since food could not be stored long enough to be transported without spoiling. These conditions led to highly modular systems in which local people were much more connected to each other than to groups farther away. In the temperate zone, the capacity to store and trade food led to exchanges governed by long-term contracts and laws, and to the development of elaborate transport systems to move food. The result was lower modularity and more central control, with institutions such as states acting to enforce laws to guard food from theft. Markets to trade grain fueled the growth of towns and cities, as well as banking and insurance.

The authors argue that because the political systems of tropical societies were highly modular, these societies were easily colonized by states from the temperate zone that had methods for setting up long-distance transport networks. The lack of long-distance exchange and long-term contracts in tropical societies inhibited their response to new technology and institutions. By contrast, in the American colonies in the temperate zone, the system that allowed for the long-term transport and sale of grain promoted a less modular and more extensive exchange of information. This led to the capacity to share technology and to set up institutions, so that initially the states were able to govern themselves and later could form a centralized federal government.

An Ecological Perspective on Modularity

Every field of biology raises interesting questions about the function of modularity. Because interactions among neurons determine what they do, neuroscientists are mapping the networks of neurons in the human brain.[33] Because groups of cells that tend to interact mostly with each other through ontogeny can form modules in body form, developmental biologists are learning how evolutionary innovation draws on regulatory modules of genes, hormones, and associated environmental conditions.[34]

By contrast, research on social networks in animals has only recently begun to consider the function of modularity and its role in the sharing of information between individuals that tend to associate with particular others. In studies of networks in animal groups, the links are some measure of spatial proximity between known individuals. These relations can persist; for example, groups of golden-crowned sparrows travel together, returning to the same sites year after year.[35] Information sharing within a group might not require familiarity; information can be shared with whatever others are nearby, even if particular individuals shift from one group to another. For example, Grevy's zebra tend to stay near particular others over the course of several months, while onagers form fluid, more temporary associations.[36] For both species, however, staying near others has the same outcome: they share information about approaching predators.

Considering the modularity of animal interaction networks raises new questions about collective animal behavior. One example comes from the study of mate choice and sexual selection. This field has largely been concerned with showing that females are choosing mates that promote their own fitness. (This work assumes, although it is not always true, that a female's fitness depends on the quality of her mate, so males compete to be chosen and females choose.[37])

However, it has proven difficult to find simple associations between mate choice and fitness. This has led to convoluted explanations for why the female sometimes fails to choose the best male: for example, she is choosing the male that successfully pretends to be best although he isn't, or she is choosing according to some feature that does not actually correspond to fitness but might be associated with something else that could correspond to fitness, and so on. Perhaps the reason it is so difficult to demonstrate that mate choice increases fitness is that often it does not; mate choice is embedded in other interactions that, along with mate choice itself, respond to changing conditions. What matters are the ecological consequences of the whole network of interactions.

Situating mate choice in the context of the modularity of interaction networks might elucidate the ecological processes important for sexual selection. For example, a long series of papers by Torgrim Breiehagen, Tore Slagsvold, and others on pied flycatchers in Scandinavian forests began with asking whether the males that mate are better than the ones that do not, thus increasing the fitness of the females that choose them.[38] They found that the males chosen more often by females neither won more fights with other males nor did any better than other males at feeding the nestlings. It turned out that males mate with more than one female at a time, that females fight with each other, and that sometimes different females have nestlings in the same nest and do not seem to prefer their own. Thus, there is an interaction network among females that involves both aggression and the sharing of parental care, and another network among males and females that involves mate choice. Which males are chosen is related to the interactions among females.

The spatial patterns in the distribution of pied flycatchers at nest sites depend on interactions among males. David Canal and his colleagues traced nest sites and breeding pairs for nineteen years in a population of pied flycatchers in central Spain.[39] They found that males arrive first and find a nest site; then females visit the nest sites of several males before deciding to mate. Mate choice by females does not depend on male traits, such as body size or color, but instead on the timing of breeding in the neighborhood. Females are more likely to choose the males that are ready with nest sites earlier than the other males nearby. These researchers suggest that the first males to find nest sites have not yet mated and are likely to devote more time to defending the nest than a later male, who may be mating for the second time in the season. Thus, female choice is also related to the network of interactions among males that influence how and when males arrange their nest sites in negotiation with the other males nearby.

These shifting interaction networks, both among males and among females, respond to changing conditions. In *The Pied Flycatcher,* Arne Lundberg and Rauno Avatalo describe the influence on the choice of nest sites of changes in weather and food availability.[40] Considering the modularity of the interaction networks of males and those of females could elucidate the ecological consequences and fitness effects of mate choice. Do certain subgroups of females tend to fight, and do they choose among only the local males? Do certain subgroups of males tend to form modules that travel together and so are competing for nest sites with only those others? How do these factors change with conditions? For example, if the weather delays the arrival of the first males to find nest sites, so that all males arrive at the same time, mate choice would shift because there would be no earliest male. If the same males tend to stay together, any condition that synchronizes their arrival would increase competition for nest sites. In general, more modular networks would constrain the choices and opportunities, making it less likely that mate choice would reflect some global measure of fitness and more likely that it would be associated with conditions that change locally.

In summary, the modularity of an interaction network influences its resilience and the speed with which the system can respond. Modularity is likely to be linked to the stability of the surroundings, the frequency and riskiness of change, and the speed with which the system must adjust.

The hypotheses presented here, linking the dynamics of the interactions that generate collective behavior with the dynamics of environments, suggest that we can learn how collective behavior works by looking at its surroundings. The next chapter outlines an ecological approach to investigating collective behavior.

7

Investigating How Collective Behavior Works

To explain how a form of collective behavior works is to identify which participants encounter which others, and how the interactions produce the outcome. An ecological approach to investigating this is to ask how the interactions respond to changing surroundings, and how these responses regulate the outcome to adjust to the current situation.

How the environment of collective behavior changes predicts how it works. How stable is the environment? How often does it change? What is the probability and cost of rupture or failure? Are resources scattered or random, or are they clustered or ephemeral?

For example, there are many forms of cancer, and their behavior differs. We know that in each form, cells evolve the capacity to act collectively as cancer in particular tissues. Tissues differ in ecological conditions such as stability, including the likelihood that spatial conditions will change or that new cells will regenerate, and the frequency of exposure to toxins. How cancer cells adjust their interactions with others is probably related to how their ancestors, healthy cells, adjusted their interactions in those tissues, with characteristic dynamics.[1] Breast cancer cells evolve in a system of ducts set to expand and change in response to cycles of reproductive hormones; colorectal cancer evolves in tiny folds subject to frequent tsunamis of peristaltic waves. An initial comparative question is about rate: How do different forms of cancer differ in the rate at which they change their interactions with each other and with healthy cells? How is this associated with the local tissue environment? Whether a form of cancer grows quickly or slowly, and what stimulates it to grow, is likely to be linked to rates of change in the environment in which it evolves. Cells that function in more stable tissue environments may draw on slower and less modular responses.

The same kind of questions apply to any natural system. There is no general recipe for choosing research questions, but there are common themes in any approach that asks how a system works in relation with what is changing around it.

The First Step: Observation

To find out what matters in the surroundings of a system, the first step is to watch it enough to learn how it changes. Observing any living entity leads us to form a narrative that pulls together strands of events, and this narrative leads to questions about what that entity is responding to. I usually have an account of what my cat is doing—deciding whether someone's lap is comfortable to settle down in, asking for food, being possessed by the urge to kill a bird outside the window. But there are still puzzling gaps in the story, those moments when I have no idea what is going on, such as when she suddenly flies down off some sleeping place in a frenzied chase down the hallway toward or away from nothing, at least nothing I can see. If I were to approach this behavior as a research project, I'd have to collect more data on when she does this; eventually I could try to see if I could do something to make her do it. (None of her toys elicit this much passion.) As in any research project, the best starting point is observation and the effort to come up with a coherent story about the relation of behavior with its surroundings.

Observation reveals patterns of change. For example, animal behavior is patterned by daily circadian rhythms: some animals are more active at night, others in the day, and there are daily temporal patterns in almost everything that animals do. Animal behavior is also patterned in space: animals repeatedly use a fixed resting place or hunting ground, contingent on conditions, such as the need to hide when predators are present, or the freedom to move when they are not. In the same way, the behavior of cells is embedded in temporal patterns, such as circadian rhythms; spatial patterns in the movement and structure of the local surroundings, such as circulation through a network of blood vessels or fibroblasts; and patterns of response to conditions, such as flowering in response to a seasonal change in temperature.

The more clearly one understands the underlying pattern, the more directly one can ask how the behavior responds to changing conditions. It would be frustrating and probably misleading to base a research project on birdsong using measures made only in early afternoon, because this would miss the birdsong that occurs in the morning. Honeybees swarm when searching for a new nest, but not otherwise; without knowing that, it would be difficult for a researcher to learn how swarming is organized.

Cells also show patterns of behavior, but they are not as easy to track as those of animals. I recently participated in a workshop of cancer researchers, aimed at formulating new research projects on the effects of interactions among cells on disease progression. The most difficult part was to identify the questions, because this requires a sense of the patterns of the behavior of cancer cells. Instead, a massive effort in elegant quantitative thinking has been dedicated to developing tools for analyzing enormous data sets that list the properties, such as genotypes, of cells. I imagined attempting to study bird behavior by putting some birds in a blender and then looking for patterns in the components of the resulting mix. It seems obvious that the first step would be to watch some birds, because one soon notices patterns in the birds' behavior, and then to form questions about what the birds do in different situations. This process works well for studying birds because we can see them. By contrast, a century of success in both molecular and cell biology, in the absence of any narratives about behavior, was based on effective methods for learning by inference about things that cannot be seen. However, the capacity to label, image, and observe cell behavior as it happens has now increased dramatically, stimulating the slower evolution of research practices. Perhaps soon there will be classes in cell behavior that return to the origins of cell biology—the observation of cells—in which students are asked to watch cells in action in their surroundings and come up with questions about their collective behavior.

This approach, based on the analysis of patterns of change detected through observation, originates in the analytic natural history that was accepted practice in the nineteenth century. In Darwin's 1881 book on earthworms, *The Formation of Vegetable Mould through the Action of Worms: With Observations on Their Habits,* he reports on a long series of his own observations and his correspondence with other English gentlemen who shared his curiosity about the behavior of earthworms.

> On January 31, 1881, after a long-continued and unusually severe frost with much snow, as soon as a thaw set in, the walks were marked with innumerable tracks. On one occasion, five tracks were counted crossing a space of only an inch square. They could sometimes be traced either to or from the mouths of the burrows in the gravel-walks, for distances between 2 or 3 up to 15 yards. I have never seen two tracks leading to the same burrow, nor is it likely, from what we shall presently see of their sense-organs, that a worm could find its way back to its burrow having once left it. They apparently leave their burrows on a voyage of discovery, and thus they find new tracks to inhabit.
>
> . . . [Their movement is] generally affected by the parasitic larvae of a fly. . . . They wander about during the day and die on the surface. After heavy rain succeeding dry weather, an astonishing number of dead worms may sometimes be seen lying on the ground. Mr Galton informs me that on one such occasion (March, 1881), the dead worms averaged one for every two and half paces in length on a walk in Hyde park, four paces in width. He counted no less than 45 dead worms in one place in a length of sixteen paces. From the facts above given, it is not probable that these

> worms could have been drowned, and if they had been drowned they would have perished in their burrows. I believe that they were already sick, and that their deaths were merely hastened by the ground being flooded.[2]

This is analytic natural history, rather than merely natural history, because Mr. Darwin and Mr. Galton made measurements of the length of the tracks (two to three to up to fifteen yards) and the density of dead worms (forty-five in an area sixteen by four paces long) to address quantitative questions. Darwin combined these observations with others on the sense organs of worms to interpret how the behavior of the worms fit with changing conditions.

These observations would not be published in a contemporary research journal because they are anecdotal. Today a rigorous investigation must include an attempt to learn if the interpretation is wrong. For the investigation of earthworms, that would begin by counting more tracks, in more conditions, to assess the variation. Enough counts must be made to see whether the counts differ more between conditions than they do within. Questions arise from considering what the earthworms are responding to and how. For example, do more dead worms appear after a rain? Are most of the dead worms parasitized? Are there more tracks, or tracks of different length, when it thaws after a snow than on a day with the same temperature when there was no snow?

After observation, the next step is to do experiments, in order to change the situation to see whether the correspondence between behavior and situation has been correctly identified. Darwin was considering the behavior of individual earthworms, not collective behavior. Experiments could make conditions warmer or wetter, or with parasitic flies kept away, in order to see if the worms respond as expected. The same questions about how collective behavior responds to changing conditions call for the same kind of experiments.

Sometimes intervention is not feasible. It would not have been easy for Mr. Galton to warm up the earthworms in Hyde Park. Similarly, if I brought a turtle ant colony to the desert, it would die. Without experiments, the best approach is comparative, asking how earthworm behavior differs on a warm day from behavior on a cold day, or how the behavior of ants in the tropics differs from the behavior of ants in the desert. If experiments are possible, then it is important to consider how the design determines what can be learned from the results.

The Next Step: Perturbation Experiments

Perturbation experiments reveal what feedback regime is at work in a system by showing how interactions among individuals respond to a change in conditions, and how those interactions then feed back to change the outcome. Feedback regimes reflect ecological conditions, which in turn suggest what kind of feedback to look for. Consider whether the default in collective behavior is likely to be to stop unless activated or to keep going unless stopped. A clue to what feedback regime is likely is whether the process uses a lot of energy relative to what it brings in; if it does, it is likely to stay inactive unless triggered. By contrast, if the activity uses little energy relative to what it brings in, it is likely to keep going unless stopped. Perturbation experiments help to identify what activates or stops the behavior.

Perturbation experiments also reveal modularity in the network of interactions among participants by asking how much a change in conditions that affects participant A also affects participant B. Ecological conditions suggest how the interactions among participants are likely to be clustered. One question is to determine how quickly the system must respond to changes in local conditions. If rapid action is more important than widespread, coordinated response, modularity is likely to be high. But if a slow response is adequate and uniformity across the system is helpful, then the interaction network may have low modularity.

Perturbation experiments are most informative when the intervention is made within the ordinary range of fluctuation or change in the surrounding system. Keeping the intervention within this range makes it possible to see how the behavior is regulated to respond to the situations it normally encounters. The use of perturbation experiments within the normal range of changing conditions contrasts with two other approaches: the knockout experiment and the factorial design.

A knockout experiment is a version of a perturbation experiment that is not within the normal range of changing conditions. It completely eliminates some element of a system. A familiar example is an experiment that provides an organism with a mutation that prevents a particular gene from functioning. The results are usually interpreted to mean that whatever is missing from the outcome is normally generated by whatever was removed. But such an experiment identifies the elements of the process, not how they are related to each other; the results do not show how the missing part is connected to the rest of the system. For example, eliminating the contribution of a particular gene has a certain outcome because of the way that gene was linked to others and to the rest of development. It is not possible to learn from the results of a knockout experiment which relations were changed in which parts of the system so as to generate the outcome; to do that would require other experiments. If the outcome of a knockout experiment is the absence of some trait or behavior, this does not mean that whatever was knocked out is the cause of that trait or behavior.

Widely used in ecology, the factorial design was inherited from agricultural research that sought to disentangle the effects of orthogonal factors. The mathematician R. A. Fisher developed statistical methods based on the general linear model, such as ANOVA and regression, to show how different factors affect an outcome. The tools for statistical analysis that apply to factorial experiments make it possible to trace the contributions of different factors in an observational study. For example, these tools can show whether a certain category of person is more likely than others to get a disease, or whether more fish are caught in El Niño years than in other years. In agricultural research, for example, when the goal in growing a new strain of tomatoes is to learn how sunlight and temperature each influence growth, it makes sense to try a range of light exposures, a range of temperatures, and the possible combinations of each. A factorial experiment could test the combined effects of different levels of light and of temperature. Maybe the tomatoes will increase linearly in size with light and temperature changes. Or there may be a statistical interaction: perhaps in low light, high temperature produces a larger tomato, but in strong light, medium temperatures are best.

Although the factorial design provides a way to predict the outcome of a process that depends on many factors, it does not reveal process or mechanism. To grow a larger, redder tomato, it is useful to know which light and temperature regime will work best, but it is not necessary to know why. To learn why an outcome occurs, a factorial experiment is a first step, but it then takes more work to figure out the process that leads to the differences associated with the level of each factor.

The goal of informative perturbation experiments is to understand the many layers of causes for any biological process. For example, it is unusual to identify a single cause in the disease progression in cancer that is sufficient to predict what will happen.[3] Instead, there are multiple causes that change over time. In collective behavior, history matters; for example, the same ecological interactions in a community of species have different outcomes depending on the sequence in which a new population enters a community.[4] The same is true for the progression of disease.

The many factors that affect natural processes are never separate. Light and temperature do not act separately to influence the growth of tomatoes; instead, the effects of these two conditions interact. (This insight is known in social science as "intersectionality."[5]) The same principle applies everywhere in nature. To learn how a tomato's growth is regulated by changing conditions, or how any behavior responds to what is around it, we must ask how different processes combine over time to create the relation of behavior and surroundings.

An experiment compares the effect of some intervention with the effect of no intervention. Because everything that happens is the result of many processes, explanations based on one factor at a time do not usually combine in a simple linear way. This is the problem with the notion of the "controlled experiment," which is based on the hope that the researcher can set things up to answer only one question about one part of a system. The controlled experiment takes the absence of some process to be equivalent to a neutral or empty state. It assumes *ceterus paribus* (everything else being equal), which sometimes morphs into the assumption that nothing else matters.

In fact, the absence of intervention or change does not guarantee there is an absence of effect. A mouse in a cage where nothing changes is just as fully in that world as it would be if it were outside. Although it does not experience the weather and all of the shifting around and daily rhythms of the natural world, its world is not empty. Even living in a cage in a room with stacks of cages, the mouse is an animal with a complete life in the world it inhabits. Stuff happened to it before, and stuff is happening to it now. If you lived in a cage, you would not be less affected by your surroundings than you are now; you would be more captive. Paul Rainey and his coauthors discuss experiments in microbial ecology from the same perspective:

> If the focus of interest were cooperative hunting in Serengeti lions, then the Serengeti Plains would be the focus of the study. For microbes, the knowledge of the appropriate environment is rarely available . . . although the value of performing studies of cooperative hunting, using caged lions and captive gazelles, would be questionable, studies of social interactions in microbes are unhesitatingly performed under conditions likely to fall well short of appropriate environmental conditions.[6]

Whenever possible I do experiments in a place where the animals live that is not imposed by me, so as to get closer to understanding responses to natural changes in situation. Often when I give talks on my work on ants, especially to people in other areas besides research in animal behavior, I am asked why I don't repeat in the lab all the experiments we have done in the field so as to find out what is "really" happening. I always marvel at the "really." It comes from the assumption that the cause of the ants' behavior is inside them, and that if we could only strip away all of the extraneous distractions, all those messy vines and lizards and wind in the tropical forest, then the essence of the ants' decisions would be visible. But this is not true. The ants in the lab will do what they do in the field in whatever way they can, but they may not engage in the behavior related to the vines, other species of ants, the lizards, the wind, the microbial community, and so on. The problem is that I don't know which behavior is related to which aspect of their surroundings, so if I studied them only in the lab, I would never know what I was missing. Similarly, experiments with cells are often done in 2D culture, but cells in two dimensions do not behave as they do in three; the signaling pathways that induce tumors to grow in 2D do not have that function in 3D, so experiments in 2D do not reveal the important 3D interactions among cells that influence glucose metabolism and regulate tumor growth.[7]

For cells, field and lab studies do not correspond to in vivo and in vitro experiments, since in vivo often entails having no opportunity to observe. Chemotherapy is an in vivo perturbation experiment that gives us no way to track the interactions of the drug with both cancer cells and healthy cells. We track the results only in measures of disease-free survival. This is like learning about how an invasive insect species influences herbivory in a forest by spraying a lot of pesticide on the forest and then evaluating the numbers of trees left a few years later from an aerial photograph. This indirect measure of the success of the pesticide reveals little about how the pesticide works in relation with all of the other processes that influence the survival of trees. Without the capacity to track interactions directly, the best option is to look for ways to create in vitro surroundings in which cells behave as they do in vivo.[8]

Perturbation Experiments with Harvester Ants

I used perturbation experiments to learn about the feedback and modularity that regulates foraging activity in harvester ants. I began with observations that led to tracing daily patterns of behavior, including foraging activity. Foraging begins early in the morning, when a wave of foragers leave the nest, and then increases to a steady state when the rate of foragers returning matches the rate at which they go out. When the day begins to get hot, at about 11:00 AM on a clear summer day, the rate at which the ants go out slows down until eventually there is a trickle of foragers returning but none going out.

From day to day, colonies adjust how many ants are performing a task and how this changes over time. To find out what regulates the temporal pattern of activity, I did experiments to perturb it. First I tried removing small numbers of ants of different tasks to see the effect on foraging and other activity. I had to begin by learning how many ants I could remove that would be enough to see an effect—but not so many that colony activity would shut down. From counts of numbers of ants engaged in each task in undisturbed colonies, made on different days that varied in conditions such as humidity, I chose to remove twenty ants of each task group, because this number was within the range of fluctuation of the counts.

I found that taking away twenty ants of one task changed the activity of ants in other tasks.[9] For example, taking away twenty nest maintenance workers led to an increase in foraging activity. This meant that the ants of different task groups were not working independently.

Then I tried increasing rather than decreasing the number of ants currently active in performing a task by changing conditions so as to recruit more ants to that task. Again, I found that changes in the conditions relevant to one task led to changes in the numbers performing another. These changes were not due to any mechanical interference, because the ants perform each task in a different place. For example, I made a mess, analogous to twigs or debris carried to the nest mound by wind or flooding, that required more nest maintenance workers to clean up. My experiment had to make enough of a mess to elicit more nest maintenance activity, but not so much that it would directly obstruct the ants performing some other task. The right amount turned out to be about fifty toothpicks placed near the nest entrance. The nest maintenance workers dragged the toothpicks to the outer edges of the nest mound and dropped them there, creating a round band of toothpicks surrounding the mound and a beautiful work of environmental art. The foragers just walked past and over the toothpicks on their way out on the trail, and on their way back into the nest.

The next set of experiments asked how changes in the allocation of ants to each task depend on ants switching from one task to another.[10] Increasing the numbers doing nest maintenance, by putting out toothpicks that the nest maintenance workers had to clear away, decreased the number of ants that were foraging. This could happen if the nest maintenance workers switched to foraging. After marking the ants of each task with a unique color of paint dabbed on their heads, I again created situations that brought more ants to a particular task. I learned that while ants can switch tasks, task switching did not explain why a change in the number of those performing one task led to a change in the number performing another. Instead, there must be interaction among ants of different task groups. The marked foragers did not switch to doing nest maintenance. The cause of the decrease in foraging activity when nest maintenance increased is that foragers respond somehow to the change in the behavior of nest maintenance workers.

These results led me to guess that ants use the rate of encounter inside the nest in deciding what task to perform, and whether to perform it at any given time. The only place that workers of different tasks meet is inside the nest, as they come inside and go out on successive trips. Foragers travel many meters away from the nest, while nest maintenance workers carry refuse out to dump ten to twenty centimeters from the nest entrance. Since the foragers cannot detect the behavior of the nest maintenance workers back at the nest, the interactions that regulate task allocation must be centralized inside the nest.

We began to investigate the encounters that influence foraging activity. In experiments manipulating the encounter rate using ant mimics (described in chapter 2), we learned that foraging activity begins in response to the rate at which foragers encounter returning patrollers. Then we asked: Once foraging activity begins, what regulates changes from hour to hour and day to day? To find out if the outgoing foragers respond to the returning foragers, I first took away the seeds of a small number of returning foragers and let the foragers return empty-mandibled. I had earlier discovered that if you tap a harvester ant lightly on the head with a small twig, it lets go of whatever it is holding in its mandibles (unless it has obtained a termite, a rare prize that a forager hangs on to no matter what). Foragers rarely return without a seed; a forager keeps searching until it finds a seed. Removing the seeds of returning foragers, and so decreasing the rate at which foragers returned with seeds, briefly decreased the rate at which foragers left the nest on their next trip.[11]

To find out if the returning foragers respond to the outgoing foragers, I tried removing returning foragers. First we removed all of the ants going out.[12] This was too drastic a perturbation; the colonies just shut down. In the next attempt, we removed the returning foragers carrying seeds for twenty minutes, picking them up off the trail outside the nest mound as they went back and putting them in a plastic box (we always let them go at the end of the morning foraging period); after twenty minutes, we stopped collecting returning foragers, so the foragers were free to return to the nest.[13] The removal of twenty minutes' worth of returning foragers sometimes shut down foraging for the day; apparently the disappearance of so many ants is outside the normal range of change in forager return rate. It's very rare for anything to stop foragers from coming in for as long as twenty minutes; maybe the trail would be shut down by several horned lizards one after another capturing foragers (I have seen two in a row a few times but never three), or by a severe weather event like a sudden downpour.

The next step was to find the minimum interval over which removing foragers would still have an effect. This turned out to be three minutes. Even over this short interval, when the rate of forager return decreased, so did the rate at which foragers went out. The rate of forager return is extremely variable, raising the question: How could such a noisy process have such a predictable outcome? Here modeling helped to specify how the rate at which ants go out can be predicted from the rate at which they return.[14] More experiments with beads coated with forager cuticular hydrocarbons and the odor of seeds, added at a rate within the normal range of variation in the rate of forager return, showed that outgoing foragers respond to both the odor of the forager and the odor of the food it carries.[15]

From there, we tried to identify the source of colony differences in how ants responded to encounters. The most recent perturbation experiments have investigated dopamine, because colonies that differ in how they regulate foraging also differ in the expression of genes involved in the metabolism of dopamine. The goal was to find a way to manipulate dopamine within the normal range. After Daniel Friedman figured out how to feed an ant a tiny amount of dopamine in a drop of water, he compared the number of foraging trips in colonies with foragers fed a neutral solution with those whose foragers had been fed the same solution with dopamine. Ants fed dopamine made more foraging trips.[16] Now more work is underway to learn how this experiment corresponds to the normal range of variation among colonies in the action of dopamine.

This series of experiments provides an example of how we can investigate collective behavior by beginning with observation and then using perturbations within the normal range of changing situations. The goal of this approach is to understand how the behavior works in relation with changing surroundings. The hypotheses that I propose here are based on the premise that there is likely to be convergent evolution, across natural systems, in this relation. The next chapter outlines the ideas that underlie this premise.

8

Evolution, Heredity, and the Causes of Behavior

The proposal outlined here is that the processes that generate collective behavior are likely to have dynamics that fit the dynamics of their situation.[1] This is based on a view of evolution as shaping dynamic response to changing surroundings. Collective behavior evolves through the evolution of the responses of individuals to interactions. These responses can be considered traits, like body form or metabolic rate. Like any trait, individual response to interactions is a heritable relation of organism and environment. One reason why collective behavior is likely to converge to use local interactions similarly in similar conditions is that an important source of innovation in evolution is new ways to adjust to change.[2] Convergence occurs because similar innovations in regulation are likely to arise in conditions that change in similar ways. A second argument for convergent evolution is adaptation: a similar solution can work well for different systems that are solving the same problem. For example, many different species of desert animals have physiological mechanisms to deal with water loss because of innovations in their evolutionary history that dealt with dry conditions, and some of these mechanisms are similar because the same ones worked well for different species. This chapter outlines the ideas about evolution, heredity, and the causes of behavior that support the plausibility of the hypotheses outlined in chapters 5 and 6.

Evolution and Heredity

Biological systems evolve changing relations with each other and their surroundings. The evolution of animals over billions of years was made possible by the evolution of flowering plants, which, through photosynthesis, put enough oxygen into the atmosphere to support large creatures that move around. This occurs on every evolutionary timescale; bacteria evolve over hours or days in response to toxins or resources; cancer cells evolve over months or years to modify their surroundings to generate a malignant tumor.

How an individual, such as a cell or an animal, participates in collective behavior is a trait or aspect of its phenotype. Asking how the interactions that generate collective behavior evolve is a form of the more general question about how any trait or phenotype evolves. This question has two parts: First, what causes phenotype and how is it inherited? And second, how does selection favor one variant over others? The familiar answer to the first question about causes is to attribute phenotype to genes. This places variation and heredity inside the organism and the ecological causes of differences among individuals in reproductive success outside the organism. In this view, the organism (or cell) is a vessel for its genes, which manufacture its phenotype, subject to the constraints of an independent and separate environment.

However, we now know that gene expression responds to conditions. This has led to a new perspective in evolutionary biology. One branch of this is "eco-evo-devo" (for "ecological evolutionary developmental" biology), linking genotype, phenotype, and environment. This branch of developmental biology examines plasticity in response to changing conditions, for example, in the development of body form.[3] Another trajectory leading to this perspective begins with Richard Lewontin's framing of organism and environment as two coupled differential equations: the environment influences what the organism can do, and the organism changes the environment.[4] Following from this, ideas about niche construction emphasize that organisms are not passive players on the environmental stage.[5] Birds or ant colonies or groups of badgers construct nests that create environmental conditions; plants secrete chemicals that change the soil in which they grow; bacteria create the conditions for other species to join with them in forming biofilms; and so on. The "modern synthesis" in evolutionary biology brought the tools of population genetics to track the process of natural selection; now the "extended synthesis" adds the insight that organism and environment influence each other.[6]

All organisms are responding to changing situations, which they in turn modify. In *Organism and Environment,* Sonia Sultan uses plant physiology and development to illustrate the relation of organism and environment working in both directions. Phenotype is always forming in relation with an environment that in turn is modified by living organisms; inside and outside are never independent. As Sultan writes: "A central feature of the organism-environment relationship is that phenotypic expression is conditioned by environmental inputs; realized phenotypes are context-dependent outcomes."[7]

It is easy to see that phenotype is dynamic in relation with conditions. But this observation contradicts the widespread preformationist view, so difficult to dispel, that phenotype is inherited because each new organism carries preexisting instructions in some inner package. While we do not fully understand the mechanisms that lead offspring to resemble their parents in phenotype, it is clear that genes are only part of the story, not all of it. How genes influence phenotype depends on what is going on around the organism. Genes are turned off and on, sometimes producing proteins and sometimes doing nothing, by processes in the surroundings. As Lewontin points out in *The Triple Helix,* heredity always arises from development; to say that offspring inherit from parents is to predict that offspring will develop as their parents did.[8] Development is noisy; two organisms with the same genes do not come out alike. This has been demonstrated many times: for example, individual cloned crayfish with the same genotype, raised in the same conditions, still differ in many aspects of phenotype, including coloration, growth, and behavior.[9] Moreover, the same genes are associated with different phenotypes in different contexts, producing a reaction norm or range of phenotypes.[10]

Galls are beautiful illustrations of heritable responses to surroundings.[11] Galls form on trees when wasps lay eggs there. A particular species of tree, in association with a particular species of wasp, develops a gall of a particular shape. This is a shape that the tree would otherwise not form. The gall is not in the genes of the wasp, nor is it in the genes of the tree. It's what happens when the wasp and the tree get together. The actions of the wasp generate actions by the tree whose outcome is a gall, which continues to change as it is occupied by different species of inhabitants.

New collective responses to conditions are a source of innovation, or variation, over the course of evolution. Variation or heterogeneity is necessary for natural selection; without variation, there is nothing to be selected. Both mathematical models, beginning with those of Sewall Wright, and empirical studies, such as those outlined by John Thompson in *Relentless Evolution*, show that the extent and distribution of variation sets up the possibilities for evolution.[12] New behavioral phenotypes are new ways of responding to surroundings. In the same way, learning and innovation in an organization happen only when its members try new and different ways of performing a task.[13]

When situations change, organisms respond, generating phenotypic change on which selection can act.[14] In *Developmental Plasticity and Evolution*, Mary Jane West-Eberhard shows that response to changing situations underlies heredity and generates variation, including the process that Waddington called "genetic accommodation," the inheritance of genetic change induced by external conditions.[15]

In this view, heritability is a result of the recurrence of a developmental process that leads in successive repetitions to similar though variable outcomes. Recurrence is a similar outcome from similar causes, and not the same as replication, which is the production of many copies from the same blueprint. As West-Eberhard writes, "The conceptual gap that should be filled by development has been filled instead with metaphors, such as genetic programming, blueprints for organisms, and gene-environment interaction." She asks: "If recurrent phenotypes are as much a product of recurrent circumstances as they are of replicated genes, how can we accept a theory of organic evolution that deals primarily with genes?"[16] A growing body of elegant work in both developmental biology and philosophy of biology, which I will not attempt to review here, shifts the view of heredity away from preformationism and toward a relation of inside and outside that leads to recurrent phenotypes from one generation to the next.[17]

The Inheritance of Behavior

It is difficult to investigate the inheritance of behavior. We cannot trace its development in the same way that we can for a flower or a foot. It is perhaps easiest to think about the inheritance of behavior as linked to body form. A cat pouncing looks like a leopard pouncing. Our cat, who lives inside, stalks silverfish. When she sees one on the bathroom floor, she hides behind the toilet, then jumps out for the kill. This behavior looks like the way a leopard hides from its prey and then pounces. The resemblance partly stems from the fact that a leopard and a refined little Russian Blue cat, who share a common ancestor, have cat bodies that make it possible to pounce. But body form does not fully explain the behavior of hiding and stalking and pouncing, which is not in the cat but instead is in the relation of what the cat does and what the prey does. A leopard cannot kill a snake in the same way it kills an impala, because the snake would not react in the same way, and if the silverfish were a cobra, my cat would no longer be around.

I do not know of any work that can trace the entire process that leads to the inheritance of any form of behavior, neither at the species level to determine how organisms develop to act more or less like others of their kind nor at the individual level to determine how variation in behavior within species persists. We cannot fully explain why offspring tend to behave like their parents. Of course, behavioral genetics has shown many instances of an association between variation in genotype and behavioral response. But we do not know the mechanisms by which the proteins produced by transcribing any particular genes lead to any particular behavior.

Sometimes the problem of explaining how behavior is inherited is presented as a merely epistemological one about what we are able to find out about the world, rather than a question about how the world actually is. For example, one approach to an attempt to distinguish internal causes from external causes is to look at the behavior of a newborn animal that has had no opportunity to learn and then conclude that whatever it does must be innate. But we can never see an animal that has not experienced the world around it, because it is always in the world, as illustrated by Gilbert Gottlieb's experiments in the 1970s with chicks.[18] He investigated imprinting, in which chicks of many bird species follow the first moving creature they see after hatching, the behavior made famous by the photo of Konrad Lorenz followed by a trail of goslings who had imprinted on him. This tendency to follow is a well-known example of behavior considered to be instinctive, because the birds do it as soon as they hatch, without any apparent opportunity to learn it. Gottlieb showed that the chicks are probably following something emitting sounds they were exposed to while they were in the egg. This demonstration reminds us that we can never see what an animal does when it is deprived of any stimuli, because there is never a time that an animal is impervious to what happens around it. Even a chick still inside the egg is listening.

Such experiments show that it is not possible to distinguish empirically between nature and nurture, between instinct and learning. But the problem is not merely epistemological, arising from the limits to what it is possible to learn by observation. It is an ontological problem as well: we cannot distinguish between nature and nurture because they are not distinct. As soon as we attempt to divide the causes of heredity into inside and outside, nature and nurture, instinct and learning, we are committed to failure.

The inability of the nature-nurture, genes-environment, inside-outside split to explain the heredity of behavior is as much a part of our everyday understanding of behavior as the duality itself. The movie *Zootopia* parodies how we imagine ancestral behavior as lying dormant inside an animal, contained in some hereditary package, ready to pop out. It uses familiar stories about the behavior of animals to tell a story about racism, which attributes inherited behavior to particular groups of people: the fox is wily, rabbits have large families, and so on. The point of the film is that our stories about inherited behavior are not true; not all predators are mean and vicious, and not all prey are kind and gentle.

Jack London's *The Call of the Wild* is a parable, perhaps about masculinity, in which behavior comes from an inner package, available to emerge when released.[19] An unruly dog lives with people who are unable to deal with his exuberant behavior. The dog ends up in the wild Yukon in the gold rush of the 1840s and is attached to various human owners, some good and some evil. Eventually he travels with his last and best owner to the wilderness, where he joins a pack of wolves. There he returns to a primal, pre-dog inner state and thus to his more fundamental true nature:

> His development (or retrogression) was rapid. His muscles became hard as iron and he grew callous to all ordinary pain. . . . Sight and scent became remarkably keen. . . . And not only did he learn by experience, but instincts long dead became alive again. The domesticated generations fell from him. In vague ways he remembered the youth of the breed, to the time the wild dogs ranged in packs through the primeval forest, and killed their meat as they ran it down. It was no task for him to learn to fight with cut and slash and the quick wolf snap. In this manner had fought forgotten ancestors. They quickened the old life within him, and the old tricks which they had stamped into the heredity of the breed were his tricks. They came to him without effort or discovery, as though they had been his always. And when, on the still cold nights, he pointed his nose at a star and howled long and wolflike, it was his ancestors, dead and dust, pointing nose at star howling down through the centuries and through him. And his cadences were their cadences, the cadences which voiced their woe and what to them was the meaning of the stillness, and the cold, and dark.[20]

It's clear that this is poetry, not biology. We know that ancestors do not howl down through the centuries. Could a dog that grew up and lived all his life in the care of people suddenly switch to being a wolf? Imagine another such transition: a person who goes to Africa and heeds the primate call of the wild by joining a troop of chimpanzees. How would the person know how to integrate into the chimpanzee social world? Even Tarzan had to grow up with the apes to learn how to swing on vines from tree to tree. Wolf-ness does not reside in a package inside the dog, ready to leap out when the dog casts off the shackles of domesticity. Instead, living and moving in packs through the forest, hunting and sharing prey, is collective behavior, carried out in relations among wolves, the forest, and the prey.

From the Inheritance of Behavior to Causes of Behavior

Explanations of the moment-to-moment causes of behavior are locked into the same split between inside and outside that characterizes heredity as genes plus environment. René Descartes's work in the early seventeenth century is emblematic of the Enlightenment view that separates the causes of our actions into mind and body.[21] In *We Have Never Been Modern*, Bruno Latour lays out how this view has entangled our thinking ever since, not just in the study of animal behavior but in philosophy and the social and natural sciences more broadly.[22] In Descartes's view, the body acts on instructions that come from the mind, the source of will. The status of mind is problematic for animals, so Descartes concluded that the animal's behavior is caused by an impulse, ultimately provided by God, that moves it. Since the impulses are made by God, who is perfect, they are good and help the animal to maintain its life.[23] Descartes's impulses correspond to our current invoking of instinct or genes, packages of instructions provided to tell organisms what to do. By analogy to Descartes's thinking, we see those instructions as made by evolution, and because we believe that evolution makes everything better, we conclude that instincts tend to help the animal do the right thing.

The idea of instinct is much older than Descartes, but no one has ever been able to explain what it is. It rests on a comparison between humans and other animals: instinct is what remains in an animal, including us, when uniquely human agency is subtracted. For Aristotle, an animal's behavior expresses its nature, and human nature includes something extra not present in animal nature: the capacity for and the use of reason and language. Instinct is a name for what we do not understand, for whatever makes animals do something that cannot be explained by rationality or intention.

The question of what causes an individual to behave the way it does is thrown around from one discipline to another, without an answer. Perhaps the actions of people can at least sometimes be explained as rational.[24] The actions of animals and other organisms, however, must be attributed to something else. Contemporary social science leaves up to biology the explanation for everything people do that is not rational, and biologists tend to pass the question along to other biologists. In research on animal behavior, the job of explaining the details of what animals do is left to neuroscientists, who in turn toss back to the realm of behavior the question of how the actions of neurons, bodies, and agency come together to produce the behavior that we see.

Questions about the internal causes of behavior get buffeted around because they are asked in a form that has no answer. Behavior is a relation of inside and outside, which change each other. Behavior specifies as well as responds to what is relevant outside it. Lewontin puts it this way in *The Triple Helix*:

> Just as there can be no organism without an environment, so there can be no environment without an organism. There is a confusion between the correct assertion that there is a physical world outside of an organism that would continue to exist in the absence of the species, and the incorrect claim that environments exist without species. The earth will precess on its axis and produce periodic glacial and interglacial ages, volcanoes will erupt, evaporation

> from oceans will result in rain and snow, independent of any living beings. But glacial streams, volcanic ash deposits, and pools of water are not environments. They are physical conditions from which environments may be built. An environment is something that surrounds or encircles, but for there to be a surrounding there must be something at the center to be surrounded. The environment of an organism is the penumbra of external conditions that are relevant to it because it has effective interactions with those aspects of the outer world.[25]

James Gibson developed the idea behind Lewontin's "penumbra" when he defined as a set of "affordances" the aspects of the environment with which the organism engages, and which the organism modifies.[26] "The affordances of the environment are what it offers the animal, what it provides or furnishes, for good or ill. . . . I mean by it something that refers to both the environment and the animal. . . . It implies the complementarity of the animal and the environment."[27] This two-way relation of organism with environment underlies the idea of niche construction mentioned earlier. Affordances change in response to the actions of the animal.

Cells have affordances too. Although the temptation to assign intention may be easier to resist with cells than with animals, it is still common practice to locate the causes of a cell's behavior inside it. For example, a huge effort in cancer research has gone into listing the genotypes of cancer cells. The hope is that particular genotypes will be shown to be associated with particular behavior, especially susceptibility to drugs: if its genotype specifies what a cell does, its genotype could specify what would kill it. Sadly, this analysis often does not work, because what a cell does depends on what is around it.

The behavior of cancer cells depends on local conditions, in ways that modulate their further growth. This insight began with Mina Bissell's work, in which she set out to explain why everyone has many cancer cells but most never lead to disease, and learned that whether a cancer cell survives depends on its relation with what is around it.[28] Recent work by many researchers shows that cancer cells die in response to chemical signals from their surroundings, including the extracellular matrix, or the "microenvironment," as well as tactile signals from other cells and relations with the activity of the immune, nervous, and circulatory systems and the microbiome.[29] Thus, the effects of chemotherapy depend on the mix of cells within a tumor, as well as the recent history of changes in that mix. The effort to specify all of this as attributes of each genotype independently is doomed to failure because the causes of a cell's behavior are not contained within its genotype.

Inside and Outside in Animal Behavior

The brief history, less than a century long, of the science of animal behavior is marked by shifts in views of what lurks inside the black box of instinct. Ideas about the causes of the behavior of animals have always been entwined with ideas about our own behavior. For example, we attribute some mix of agency, intention, cognition, and emotion to animals by analogy with ourselves. Ideas about the causes of animal behavior then carry over to the personification of other living beings: for example, cells can appear to act like individual selves, akin to animals, which in turn look something like people.

The comparative psychology of the late nineteenth and early twentieth centuries, which investigated mental processes in humans and animals, branched off into psychology focused on humans; animals represented steps on the way to humans. The current science of animal behavior took shape as ethology in the 1930s and '40s, mostly adhering to the behaviorist position that the scientific study of animals should leave aside any attribution of mental processes.

Ethologists such as Lloyd Morgan, Julian Huxley, Nikolaas Tinbergen, and Konrad Lorenz were explicitly interested in animal behavior as distinct from human behavior. Many were ornithologists, whose ideas about behavior were based on observations of birds, especially on how birds communicate with each other. The herring gull chick tells its mother that it is hungry by pecking on the red spot on her beak. One male tells another that it is large enough to win a fight without actually fighting.

For these early ethologists, behavior consisted of preformed strings or nuggets that were essentially static, called "fixed action patterns." The initial project was to list all of the fixed action patterns of an individual of a species and all of the stimuli associated with those movements. Lorenz's hydraulic model, sometimes known as the "flush toilet" model, pictures how internal motivation, or the drive causing the animal to act, is triggered by a stimulus from outside the animal. There is a tank full of motivation, and when some stimulus pulls down on a lever, the motivation flows out, causing the animal to perform a particular set of movements. Instinct is this arrangement, which links behavior to motivation in the presence of an external stimulus. It tells the herring gull to move its head in a certain way when events provoke anxiety, or to feed a nestling when pecked on the beak. The outside stimuli are independent of the animal. Neither the behavior of the herring gull nor the surroundings can change the way its hydraulics are set up.

This notion of the animal's behavior as fixed, each action prompted by an external trigger, is not easy to sustain when watching an animal—at least, I've never been able to do it. Behavior always seems more variable and more flexible than a string of fixed action patterns. Researchers have to settle somewhere in between this form of explanation and what they are actually seeing, and the difficulty of bridging this uncomfortable gap is what allows vague ideas about instinct to persist. Maybe Konrad Lorenz saw his geese perform a series of acts that were each powered by little hydraulic tanks in their brains, but it is clear from Tinbergen's descriptions of gull behavior in *The Herring Gull's World* that he began by understanding the flow of gulls' responses to each other and only afterwards packaged them as fixed action patterns.[30]

In World War II, the new technology of computers and control reinforced the idea of animals as cyborgs, or animated machines, ideas that were familiar since the ancient world and were captured in Lorenz's hydraulic model.[31] Haraway describes how information theory came into the study of animal behavior as cybernetics began to pervade engineering.[32] Behavior was a system of signals. This was an easy progression from ethology; the fixed action patterns became switches to stimulate a communication event, in the same way that pressing a key on your computer can stimulate an email message. In this view, each piece of behavior is still discrete and bounded, and the initial state and the rule that links the switch and the action fully predict behavior.

Researchers in animal behavior have long recognized that the divide between instinct and learning does not work.[33] Aubrey Manning and Marian Stamp Dawkins's textbook on animal behavior sidesteps the question this way: "We seem to have come up with a clear dichotomy—instinct (or nature) in which adaptation occurs over generations by selection, or learning (nurture) where adaptation occurs within the lifetime of an individual. . . . Studies of the way in which behavior develops do not support such a distinction, and suggest that the categories themselves are inadequate."[34] If the categories are inadequate, why do we keep using them, like Sisyphus pushing the boulder up the hill only to see it fall down again?

What Causes a Harvester Ant to Forage?

How then can we understand the causes of behavior without attempting to parse them into separate internal and external bins? To learn about collective behavior in any system, whether ants or cells or trees or dogs, is to trace the relations that link interactions with changing situations. It starts by describing behavior as an ongoing relation with changing surroundings. Although it is a recent innovation to speak of plants as "behaving," this describes the flexibility in the plant's growth, form, and chemistry.[35] Perhaps because plants behave slowly, and individual plants of the same species look alike to most of us, we do not try to divide their behavior up into static and discrete acts carried out by unique individuals independently of their surroundings. Sunflowers turn toward the light, vines crawl along branches, leaves put out defensive chemicals when they have been eaten by caterpillars, and trees grow out of cracks in rocks. In the same way, the behavior of an animal or a cell is a response to the conditions it experiences.

Let's take a specific example. What causes a harvester ant to go out of the nest, search for a seed, and bring it back?

First, whether a harvester ant is a forager depends on its physiology and gene expression, which are related to what it did and who it met the day before, who it meets today, and the situations it encounters. When it takes up a particular task, there is a shift in the daily rhythm of the expression of genes associated with moving around.[36] In a daily cycle familiar to parents of newborns, it seems that the genes associated with activity for the ants feeding the larvae are transcribing in the middle of the night, while the peak for foragers is in the morning, which is when they go out to forage.

It is an open question whether daily temporal patterns in gene expression produce daily rhythms in activity, or whether activity induces changes in gene expression, or both. Certainly the activity of harvester ant foragers depends on external cues like temperature and sunlight. If sunlight is the switch that turns on the genes that turn on the ant, then there would have to be a new set of switches that operate whenever the ant changes task to be active at a different time, or to perform a different task in response to encounters with other ants. Behavior changes much more rapidly than gene expression does, and once genes are expressed to produce proteins, there are still many steps, mostly unknown, that link the proteins to behavior, all also influenced by conditions.

A forager is not active all the time. Whatever process makes a harvester ant into a forager today and likely to forage tomorrow does not determine whether it is foraging right now. Whether an ant forages depends first on whether it is a day when the rest of the colony forages at all. Early in the morning, some process, perhaps stimulated by light in the nest entrance or the warmer temperatures after sunrise, moves the foragers up from the deeper nest into the chamber just inside the nest entrance, ready to go out. The first ants to go out are the patrollers, and their safe return leads to encounters with foragers. When patrollers return at a rate of about ten per second or higher, the first wave of foragers leaves the nest.[37] On some days a colony does not leave the nest at all; when we look inside the nest of an inactive colony with a fiber-optic videoscope, there are no ants in the entrance chamber. The foragers have not come up from the deeper nest.

Once the foragers have come up to the entrance chamber and the patrollers have gone out and returned to the entrance chamber, the first wave of foragers leaves the nest. Their direction depends on the patrollers' choice of foraging trails for that day, apparently mediated by encounters between patrollers of neighboring colonies that help to prevent overlap in their respective foraging trails.[38]

Each forager goes out on many trips a day. When it comes back from a trip, it drops its seed in the entrance chamber and then waits around or goes back down to the deeper nest. Whether a forager in the entrance chamber leaves on its next trip depends on its rate of encounters, over the course of about thirty seconds, with returning foragers carrying food.[39] This system uses positive feedback, which increases foraging activity when more food is available, as described in chapter 2. The rate at which the forager meets another, and whether it goes out again, depends on how quickly the foragers return with food, as well as on the local density of ants, which in turn depends in part on the shape of the nest chamber and tunnels.[40] If the forager does not meet any other foragers for a while, about five minutes, it gives up on waiting to be stimulated by returning foragers and goes down to the deeper nest.[41]

The harvester ant forager's decision to leave the nest also depends on the current humidity. Ants lose water to evaporation when the air is dry, and in dry conditions foragers are less likely to go outside. Inside, the nest remains cool and humid throughout the day, even as it gets hotter and drier outside, so a forager can assess the humidity outside only from its experience on its previous trip.[42] On dry days, and especially in some colonies, a forager is less likely to leave the nest in response to encounters with other foragers.[43] Dopamine plays a role in the influence of humidity on a harvester ant forager's decision to leave the nest.[44] It seems that dopamine overcomes the ant's reluctance to forage because of the risk of water loss.

Once the forager is out of the nest and on the trail, it joins a diffuse stream of ants traveling away from the nest. On each of its trips that day it tends to leave the foraging trail at about the same place.[45] Perhaps the forager can remember, during the short interval of a minute or two it waits inside the nest between trips, how long it walked on its last trip before it left the trail. Once it leaves the trail at its highway exit for the day, it searches for seeds, either on the surface or buried in the soil. I have seen searching foragers stop, dig almost a centimeter down in the soil, and come out—I'd like to say "triumphantly"—holding a *Plantago* seed. As soon as the forager finds food, it turns back to the trail. The return journey may involve cues from the position of the sun. It also requires interactions with other ants; the returning ant travels down the stream of outgoing foragers, occasionally touching antennae to their bodies and thus confirming that they are nestmates. In experiments in which we removed the outgoing ants, the returning foragers seemed to get lost. The forager uses contact with outgoing foragers, as well as a chemical gradient of the colony's odor on the nest mound, to find its way back to the nest.[46]

Thus, the forager goes out in response to others, travels along with others, and comes back with others. Whether it leaves the nest on the next trip depends on processes both inside and outside the ant, on many different temporal and spatial scales. These include daily temporal cycles in its gene expression: the current temperature and humidity, which depend on how many days have gone by since it rained and how much it rained; how much its cuticular hydrocarbons allow water to evaporate out; its dopamine neurophysiology; the process involving other ants that roused the forager to move up from the deeper nest into the entrance chamber that morning; whether whatever happened to the patrollers outside allowed them to get back safely to encounter the forager; how much it rained the previous season, which influences what seeds are available that day; where the seeds are and how quickly other foragers find them and return;

whether the forager met returning foragers with food over the past thirty seconds; the shape of the entrance chamber of the nest, which constrains the flow of ants and thus influences the probability that an outgoing forager meets a returning one; and much more.

The forager is participating in collective behavior through its encounters with other ants, and those encounters are related to all of their affordances, the "penumbra" created by ants responding in their world. As many have pointed out, it does not make sense to divide all of this into a separate inside and outside.[47] I do not know all the causes of a forager's decision to leave the nest on the next trip, but I doubt that it is getting instructions from a package inside, whether we call it the voice of God, genes, instinct, a program, or wiring. The inside causes are not any more important than the outside ones. It's impossible to disentangle the daily rhythm of the expression of the genes that are making proteins when the forager is active, the activity of its dopamine neurons, its perception of the current humidity, how much water is in the air, and its response to the smell of other ants, responses that influence how it stimulates other ants and, in turn, contribute to its experience. All of this works together.

The causes of the ant's behavior are woven into the relations linking the ant, the place it is searching, the other ants, the seeds, and the humidity in a web that widens as we learn more. This amounts to a more elaborate and detailed version of Aristotle's idea that the animal is expressing its nature. The important point is that its nature is not inside it, but instead in how it reacts to and changes its situation.

The view of evolution outlined in this chapter is that the causes and heredity of traits, such as behavior, are generated by the relation of inside and outside, not contained in a packet of instructions carried inside. This perspective on the evolution of collective behavior is emerging in every field of biology and underlies the hypothesis that the dynamics of collective behavior reflect adaptation to the dynamics of the environment in which the behavior evolves. To summarize: The dynamics of collective behavior are generated by how individuals interact to adjust collective outcomes to changing conditions. Selection for collective behavior is selection on individuals for how they interact to contribute to collective outcomes. Evolution shapes individual participation because of the ecological consequences of the collective outcomes for the individuals.

This approach to investigating collective behavior differs from the prevailing one in behavioral ecology, which frames collective behavior as a set of choices by individuals whose interests are independent of those of the collective. The next chapter explains why an approach asking how interactions allow collective adjustment to changing situations is more productive, and more biologically realistic, than an approach that presumes that the interests of the individual and collective are in conflict.

9

From the Collective to the Individual

The Strategic Individual

Hypotheses for how collective behavior fits changing situations can help us learn how collective behavior works by focusing our efforts on looking for similar processes in similar ecological situations. This is an evolutionary hypothesis: because collective behavior adjusts to changing conditions in ways that have ecological consequences for its participants, selection acts on how individuals produce collective behavior.

Questions about how selection acts on individuals and groups have led to decades of debate. Elisabeth Lloyd resolves the issue with the straightforward argument that, when collective behavior has ecological consequences for both the individuals and the group, selection can shape collective behavior through individual interactions.[1] The work of Michael Wade and others shows that group selection can occur and is not necessarily in conflict with selection on individuals.[2] The interests of the individuals and of the group may be distinct and may combine in different ways, so that selection may act on both individuals and groups.[3]

This view of the evolution of collective behavior, which I draw on here, does not assume that the interests of the individual are pitted against those of the group. By contrast, the premise of most work on the evolution of collective behavior is that individual interest and the collective good are opposed.

I will outline two ways in which the premise that the interests of the individual conflict with those of the collective impedes investigation of the evolution of collective behavior. First, the predictions of this approach are generally not supported by the data and are based on quantities that are impossible to measure. It seems to me more efficient to start out with hypotheses that are likely to be confirmed, about behavior that we can observe. Second, this approach supposes an evolutionary trajectory that we know never happened, in which selection acted on variation among independent individuals in how they chose to participate in the group. Of course there never were any such independent individuals; this trajectory is merely metaphorical. It would make more sense to investigate evolution as it could actually have happened, or as it is currently happening.

In *Greek Political Thought,* Ryan Balot outlines a trajectory from the views of some philosophers of ancient Greece to the modern version of the strategic individual.[4] The Greek philosophers saw a natural trajectory from animals to humans to gods. In their view, humans create a political structure because they have justice, honor, and virtue, by means of language. They thought that these capacities make humans better than the animals, who can get by with simpler and less admirable behavior using their natures or instincts. In the fifteenth century, Niccolò Machiavelli introduced a different, instrumentalist view of humans in which the individual seeks to maximize his personal gain. From this point on, the animal as strategic individual could be used to justify why we are as bad as the animals rather than to explain why we are better.

Machiavelli's trajectory leads to Thomas Hobbes's *Leviathan* in the seventeenth century, in which he argued that each man sacrifices his autonomy to the state so as to be protected from the greed and violence of others.[5] About a century later, Adam Smith described how strategic individuals can form a society that benefits all, by engaging in contracts that allow each man to trade some of his resources for the opportunity to gain different resources from others.[6] Garrett

Hardin's "Tragedy of the Commons," published in 1968, takes the opposite view, arguing that when everyone can take as much as they want, individual interest acts against the collective good.[7] These economic narratives pervade our accounts of living systems. *Homo economicus*, the rational decision-maker who decides what to buy so as to maximize his profit, became *Animalis economicus*, who decides what to do so as to maximize the fitness, or capacity to replicate, of its genes. I do not mean to dismiss the human capacity for greed and love of power, but to argue that the human pursuit of self-interest does not explain the evolution of collective behavior in natural systems.

The idea that individual and collective interests are not always opposed in biological systems has emerged many times. Pyotr Kropotkin, a Russian aristocrat with a strong interest in biology, brought together his observations of animals in the Siberian steppe with his sympathy for anarchism as a means toward a society without central control. In his book *Mutual Aid, published in 1920, Kropotkin* points out that animals do not always act antagonistically toward each other and instead act in ways that benefit the group.[8] Allee's work in the 1930s showed that some animals, including goldfish, mice, and flour beetles, are more likely to survive amid a high density of the same species, leading him to suggest, as Kropotkin did, that animals help each other for the good of the species.[9] Similarly, V. C. Wynne-Edwards argued that animals regulate resource use so as to maintain the viability of populations.[10]

George Williams, in *Adaptation and Natural Selection*, published in 1966, scorned the idea that animals act for the collective good.[11] He thought this idea showed a foolish disregard for the conflict between individual and group interests: Why would selection act on reproducing individuals to act for the good of the group? Williams's book set the stage for Richard Dawkins's *The Selfish Gene* in 1976, which casts the economic conflict between individuals and groups as biological, with offspring rather than money the currency that individuals seek to maximize.[12] Dawkins personified genes, conceiving them as "selfish" packages that cause behavior and the rest of phenotype. In his account, not only do genes tell everything else what to do, but they do so with a motive, namely, making more of themselves. Patrick Bateson responded to this by pointing out that to say that genes are ways of making more genes is like saying that a bird's nest is a way of making more nests.[13] In other words, genes participate in the process that makes more genes, just as nests participate in the process that makes more nests, but genes do not have more fundamental power or control than nests do.

In *Primate Visions*, Haraway uses the history of research on primates to trace the transformation of animals into strategic economic actors.[14] When primate research began in the first part of the twentieth century, primate behavior was portrayed as a version of the nuclear family headed by Man the Hunter who provides for his passive mate and offspring. As women became researchers, first as assistants to male professors and eventually as independent researchers, primate studies began to show the role of the female primate as more active and consequential in the primate family, and emphasis shifted to the viability of the family.

Then the theme of the nuclear family began to meld with an economic view of individual profit. The mechanical animal began to look out for itself alone, propelled by its selfish genes. This view disrupts the cozy nuclear family, whose interests are joined. An example of this transition from family to the strategic individual is the study of the dunnock, a "drab and homely bird of English forests," by Nick Davies and his colleagues. Using genetic variation to find out who the birds were mating with, they discovered that although a pair seem to mate for life, in fact females mate with many males other than their partners. This changes the story about dunnocks as models of harmonious nuclear families. Davies writes in *The Social Life of the Dunnock*:

> The Reverend F. O. Morris (1856) encouraged his parishioners to emulate the humble life of the dunnock, or hedge sparrow *Prunella modularis*. . . . A change in focus from casual observation of the species to detailed recording of the lives of individuals reveals that this Garden of Eden view of dunnock life is very much mistaken. The Reverend Morris's recommendation turns out to be unfortunate:

> we now know that the dunnock belies its dull appearance, having bizarre sexual behaviour and an extraordinarily variable mating system. Had his congregation followed suit, there would have been chaos in the parish.[15]

Here Davies contrasts the story of the cozy family with another in which females and males all seek to maximize their own fitness. This new story is based on a conflict between personal gain and the collective good or gains by others. Of course, a society in which everyone works exclusively for their own advantage would not function. Without some participation in collective behavior, there would not even be a parish, chaotic or otherwise.

The field of behavioral ecology devoted its first few decades, beginning in the 1980s, to testing economic models that predicted how each individual would work to optimize its own reproductive success. This project asks how benefits for individuals, for example, in obtaining food or fighting off enemies, exceed expenditures in effort and opportunity. Similar calculations frame social behavior within a family as a set of individual choices, measuring the benefits of choosing mates of one type relative to another, or of providing food or protection to offspring. Applied to collective behavior, this project asks how many resources or how much fitness an individual could obtain in return for the price it pays by being in the group. In other words, it asks what is the marginal utility of joining the group.

Predictions about individual gain derived from the premise that individual and group interest are opposed are rarely confirmed by the data. John Fanshawe and Clare Fitzgibbon's study of wild dogs provides an example.[16] The study asked how the size of a hunting group was related to hunting success. Wild dogs hunt prey such as gazelles and wildebeest. A single dog can kill a young gazelle, but it takes a group of dogs to bring down an adult wildebeest. Fanshawe and Fitzgibbon calculated the meat per individual for hunting groups of different sizes. They found that the group size does not maximize the amount of meat per dog. The same number of dogs might hunt a gazelle as a wildebeest, although the gazelle provides much less meat per dog.

In *The Logic of Collective Action*, the economist Mancur Olson calculates the impact of group size on the amount of each individual's share of the collective good.[17] For the group of wild dogs, the collective good was meat. Olson considers companies and labor unions, for which the collective good translates into money. He shows that the effect of group size on each individual's share of the collective good depends on how gain or benefit and cost are each proportional to group size. Gain has to increase faster than cost, so that the larger the group the smaller the cost, to make up for having to divide up the gain. Another way to say this is that as group size increases, the marginal cost for each individual of more of the collective good has to be partitioned in exactly the same proportion as the additional benefits. Olson argues that, though a group could be large enough to create enough benefits that individuals get more, this requirement is rarely met in human organizations. Instead, as a group increases in size but does not become large enough, people get a smaller share of the collective good.

This is what happened, in Fanshawe and Fitzgerald's account, for the wild dogs. Joining the group did not always increase the dog's meat intake. Innumerable studies show the same: animals do not always behave strategically. Fanshawe and Fitzgibbon's study found another consequence of hunting in groups and gathering to share the meat, beyond the amount of meat obtained per animal: the larger the group of wild dogs feeding together, the longer they can fend off the hyena that come to feed on the dogs' kill. Perhaps the collective behavior that allows a group of dogs to defend their kill from hyena makes it worthwhile for each dog to get a little less meat. It seems unlikely that each additional dog deters hyena by an equal amount; instead groups of dogs engage with groups of hyena in ways that differ with the group size of each species, and so the per-dog meat index may not scale with the effect of group size on deterring hyena. Other benefits of functioning as a group, besides deterring hyena, that may be more important than meat per dog per hunt include opportunities for puppies to practice hunting with others, protection from predators, finding water sources together, and the oxytocin rush from companionship.

Another study of wild dogs suggests that group size is related to the collective process of dispersal and forming new groups. Rosemary Groom and her colleagues compared the pack sizes of wild dogs in the Save Valley Conservancy in Zimbabwe during a time when there were few lions, only six in a large reserve, and later when the lion population had increased to about 120.[18] Lions and wild dogs eat the same ungulate species. When the lion density was higher, the lions occupied the areas with more prey and the wild dogs retreated to rougher ground in areas with fewer prey and fewer denning sites. Pup survival diminished, apparently because of the lack of prey, and pack size decreased. The size of hunting groups may depend on broader collective processes, such as interactions among dispersing packs that are all responding to hunger, or the rate at which they encounter prey. In turn, the food supply shifts the age structure of packs and thus interactions within the pack.

Asking whether a dog chooses to join a hunting group if it gets enough meat by doing so evokes an evolutionary scenario like the following. First there was heritable variation among dogs in willingness to join a group; some dogs, rather than hunting on their own, joined with other dogs, and this propensity to join a group was inherited from parents by their offspring. Then the dogs that were more sociable got more food and were able to produce more sociable offspring. One problem with this scenario is that the ancestors of wild dogs lived and hunted in groups. There was no ancestral solitary canid that weighed the costs and benefits of sharing meat with others. Instead, there must have been variation among dogs in how they interacted with each other in groups in response to what was happening around them. Earlier forms of collective behavior evolved into the ones we see, depending on the ecological consequences of the relations among dogs in groups, not as individual decisions about profit made independently of being in a group. Looking at the benefits of being in a group might help us learn about what is happening now, but it does not explain how collective behavior evolved.

In general, when animals do not conform to the prediction that their behavior prioritizes individual benefits over the collective good, there are two possibilities. Either the behavior just does not maximize individual benefit, or the collective provides other benefits to individuals that are not captured by the model. In fact, the benefits of collective behavior are always context-dependent and historically contingent. This is why the most direct approach to learning about the benefits of participating in collective behavior is to learn how the system responds to changing conditions. To try to explain what the individual should do to maximize its profit in each separate transaction and then only later, when that fails, look at what is going on around the individual, is to begin with a detour. The same objection has been made to the analogous ideas in economics that were the origin of this approach in evolutionary biology. Much of contemporary political economy investigates the layers of social structure that surround the economic choices of individuals.[19]

Early on, the response from behavioral ecology when empirical tests failed to show that animals behave optimally to promote their own interest or fitness was to argue that any failure to confirm optimality just meant that the right test had not been made; the behavior was optimal for some other reason. This response relies on the adaptationist assumption that any trait we see, including behavioral ones, must have evolved because that trait was selected for. This is the view that Williams criticized in *Adaptation and Natural Selection* and Stephen Jay Gould and Richard Lewontin objected to in "The Spandrels of San Marco."[20] The position that any failure to find optimality merely shows that the test was incorrect is, of course, logically unsatisfying, but it also raises a heuristic problem: it favors looking to confirm what is believed to be true rather than investigating what is actually happening.

It is sometimes suggested that even if it is not true that animals generally evolve to promote their own individual interests above those of the collective, this perspective is useful because it helps to stimulate and frame research. As usual, Lewontin had the last word on this idea:

> That a theoretical formulation is desirable because it makes it easier and more efficient to write more articles and books giving simple

> explanations for phenomena that are complex and diverse seems a strange justification for work that claims to be scientific. It confuses "understanding" in the weak sense of making coherent and comprehensible statements about the real world with "understanding" that means making correct statements about nature. It makes the investigation of material nature into an intellectual game, disarming us in our struggle to maintain science against mysticism.[21]

Not Reproducing

The first problem with the attempt to explain collective behavior by showing that individual benefits outweigh the cost of participation is that the predictions of this approach, as discussed earlier, are rarely confirmed. The second issue, related to the first, is that this approach presumes that the costs and benefits to the individual can be measured, while in practice this is usually impossible. The currency is some proxy for fitness, which itself is a slippery concept, and there is always a question whether the proxy, such as more food, more mates, or more territory, really is associated with fitness. More fundamentally, the fitness of any individual is entangled with the fitness of others. Identifying the benefits that a participant in collective behavior receives and the costs that it pays is like trying to add up how each finger adds to the fitness of an elbow, and what it costs the finger to be attached to an arm.

Even in explanations of the collective behavior of people, who are at least in principle capable of rational behavior, it is not easy to parse out for any individual the costs and benefits. Socrates famously chose to obey an unjust death sentence and take the poison that killed him because he felt he owed obedience to the laws that had provided him with so many benefits during his lifetime.[22] Arguing that the high cost, death, was balanced by the benefits of citizenship, Socrates did not seem concerned about the effects of his disgrace on the prospects of his offspring and thus on his own reproductive success. But there is no realistic assignment of value that could compare the cost of death relative to the benefits of an Athenian education.

Explanations of the evolution of colony life in social insects show the difficulty of measuring the costs and benefits of individual contributions to the collective good. In social insect colonies, one or a few females reproduce, while other females do not. This is framed as a trade-off between reproducing and doing other work. Not reproducing is "altruism," because it imposes a cost on the individual worker and also brings some benefit for the others from whatever else that worker does.

In the 1960s, W. D. Hamilton posed the question about the evolution of altruism by imagining that the workers have a gene for not reproducing.[23] He asked how such a gene could possibly persist. Shouldn't it die out, since it inhibits reproduction? His answer drew on the peculiar haplodiploid genetic system of ants, bees, and wasps. *Hymenopteran* males are haploid, with only one copy of each chromosome, while females, like humans, have two copies and are diploid. The result is that, if the queen mates only once, workers are more related to their sisters than they would be to their daughters. Thus, Hamilton reasoned, if a worker's mother gave her the gene for not reproducing, then that worker contributes more to the spread of that gene by helping her mother make sisters than she would if she made her own daughters.

It is easy to see that not reproducing imposes a cost, if the currency is the amount of reproduction. But there is no way to measure the benefit to itself and others that the worker provides by not reproducing. How much does it benefit an ant to bring back food and contribute it to the common store of the colony, and how much does it benefit others? How could we distinguish the portion of the ant's benefits from what it eats itself from the portion that goes to help raise the ant's sisters? How do we measure the contributions to their own and to the collective good of each worker that helps to build a nest or carry out the garbage? Some workers mostly hang around doing nothing. They may function as reserves, ready to work if needed, or as receptacles that store food or water. If they are giving up reproduction by being sterile, but also not contributing anything to others, are they altruistic? Although a great deal of effort has gone into measuring the costs of not reproducing, attempting to measure instead the gains of fitness for each worker has not led to any insight into how colonies operate collectively.

Since Hamilton suggested in the early 1960s how the haplodiploid system of Hymenoptera could promote the evolution of worker sterility through kin selection, there has been enormous progress in the genetic methods needed to measure paternity. We now know that in many species of social insects the queen mates more than once, so Hamilton's suggestion, while a brilliant idea, and one consistent with the information available at the time, does not explain how worker sterility persists, because workers are not more related to their sisters than they would be to their own offspring.

West-Eberhard and others have offered an alternative explanation that is not based on calculating costs and benefits to individuals. They compared subsocial species of wasps in which some workers reproduce and others do not, depending on conditions.[24] These subsocial species of wasps are the ancestors of the eusocial ones in which workers never reproduce. Wasps as a group are the ancestors of the ants, who also have sterile workers. In this account, worker sterility in wasps, and then in ants, is a deletion of part of the behavior of the ancestral subsocial female wasps, who show a range of ovarian development, from none to full, regulated by seasonal changes in conditions and by interactions among females. What has not been deleted over evolutionary time is that a wasp takes care of larvae it encounters. While in some individuals the responses to seasonal conditions and to interactions that stimulate ovarian development and laying eggs have been lost, wasps have not lost the response to larvae that stimulates feeding. In this case, when larvae do not develop over the winter, there will be colonies in which different life stages live together. This account of the origin of eusocial colonies does not require selection for a gene for not reproducing.

Altruistic or helping behavior in birds, when birds care for the young of other parents in a group, raises the same difficulty in measuring costs and benefits to the individual independently of the group. An excellent review by Walter Koenig and others outlines the many attempts, not always successful, to show that helping behavior benefits the helper.[25] When the helper turns out to be related to the parents, as sometimes happens, helping can be explained as furthering the fitness of the helper's relatives. But when the helper is unrelated, that explanation does not work. Maybe the helper learns to be a better parent, or maybe hanging around makes the helper more likely to inherit the territory of the parents, or maybe having the helper there has some other benefit, such as defending the whole group's territory and thus giving the parents time off from fighting that they can use to get more food. Attempts to demonstrate these advantages to the helper have often failed, however, because it is not possible to disentangle the effect of the helper from other aspects of the situation. For example, when there are more helpers in places where there is more food, it is not possible to separate how much each individual is benefiting directly from the food and how much from the behavior of the helpers.

The question of why some individual birds do not reproduce may have an answer similar to the one for wasps; just as wasps tend larvae even when they do not develop the capacity to reproduce, adult birds protect and feed juveniles even if they do not mate and produce offspring themselves. More generally, the more interesting questions about who reproduces are about how individuals shift their roles in relation with changing situations. Recent work shows that cooperative breeding in birds is related to environmental stability.[26] For example, Dustin Rubenstein and his colleagues show that, in a comparison of breeding behavior in forty-five species of African starlings, cooperative breeding was more likely in species that lived in areas with more variable rainfall, as it provided opportunities for individuals to keep breeding in drier years. Within one species, survival was higher in dry years for females in larger groups.[27] Studies such as these provide insight into the ecological consequences of the behavior of the group of birds, showing how the group behavior benefits the individuals that engage in it, and how selection can favor the genetic processes that are associated with the response.

Mutualism

In mutualistic relations among species, each species uses something that is generated by another species. (Interestingly, participants in mutualisms are often species that work collectively, such as ants, perhaps because collective behavior uses sustained patterns of frequent interactions within a species that can extend across species.[28]) Often this relation is depicted as a transaction between strategic individuals, each providing a service to the other and exacting a price, or gaining a benefit from the other that is equivalent to the cost. This characterization is misleading because the costs and benefits depend on a changing situation in which relations are continuing to evolve. Mutualisms are not static but constantly evolving.[29]

Beth Pringle's work on the mutualistic relation of *Cordia* trees, *Azteca* ants, and scale insects, along a gradient from Mexico into Costa Rica, provides an example of the growing understanding that mutualisms continuously evolve.[30] The *Cordia* trees grow cavities in which the ants live, and their sap feeds the scales that live in the same cavities. The ants feed on the sugary excretions of the scale insects. Thus, the plants feed and house both the scales and the ants, and the ants defend the plants from the herbivores, mostly caterpillars, that eat leaves.

The ants are currently evolving in relation with the amount of water available to the plants. In Mexico the plant loses its leaves in the dry season; once the rainy season begins, new leaves sprout immediately. These new leaves produce much of the plant's annual nutrition, so defense of the new leaves from herbivores, provided by the ants, is especially important. In Costa Rica, rainfall is more abundant and more consistent over time than it is in the Mexican dry forest, and the *Cordia* trees there are not deciduous. Thus, defense from herbivores is not as crucial in the wetter climate of Costa Rica as it is in the seasonally dry forest of Mexico. Pringle found that the trees in Mexico produce more carbohydrates in their sap to feed the scales, which in turn goes to the ants defending the new leaves, than they do in Costa Rica, where they have leaves year-round. The ants all along this gradient are of the same species, but their behavior differs along the rainfall gradient. The ants in Mexico, better fed than their Costa Rican relatives by the extra resources that the plants there give to the scales, are faster at attacking herbivores. The relation among these species is evolving through the ecological consequences for all three.

Considering each species in a mutualism as a strategic individual frames the relation as one of potential conflict. For example, one partner in the mutualism might cheat, taking more than it provides. Some have suggested that in response to cheating, mutualist species evolve sanctions. In this scenario, the mutualistic relation evolves first, and then one side evolves some practice to punish the other partner if it does not do its share.

Megan Frederickson argues that there is little evidence that sanctions against cheating have contributed to the evolution of mutualisms.[31] Instead, the mutual adjustment that could be characterized as a punishment for cheating arose prior to the evolution of the mutualism itself and was part of the process that allowed the mutualism to evolve further.

Frederickson provides a compelling example of an evolutionary exchange that defies characterization as a sanction against cheating. Yucca moths benefit yuccas by pollinating their flowers, but they are also parasites; they lay eggs in the yucca flowers, where the larvae eat the growing seeds. Yucca plants sometimes abort the flowers that are parasitized by moths if the flower does not get enough pollen. Aborting the flowers was interpreted as a sanction against cheating. In this view, the yucca is punishing the moth for not doing its job, by taking away the site where the moth lays eggs. Frederickson shows that instead, yuccas first evolved to abort flowers that get insufficient pollen, before the yucca moth evolved the practice of laying its eggs in yucca flowers. In fact, aborting flowers without pollen promoted the evolution of the mutualistic relation with yucca moths as pollinators, because it selected for moths that bring more pollen into the flower that provides them with a place to lay eggs. Phylogenetic studies show that the moths evolved tentacles that pack up pollen to carry from flower to flower only after they evolved to lay eggs in the flowers. Thus, they became better pollinators after they became seed parasites. The plants and the moths both evolved in relation with each other.

The more we learn about the evolution of mutualisms, the more difficult it becomes to frame the process as a sum of costs and benefits to independent players. Instead, we see a dynamic balance of relations among the species that sets up how they operate and evolve together, with ecological consequences for all the partners collectively.

Investigating Current Natural Selection on Collective Behavior

When we can see the ecological consequences of variation in collective behavior, we can find out whether natural selection is currently shaping it. Variation in the ways that individuals respond to conditions is the raw material for selection.[32] For collective behavior, these are reaction norms in how individuals respond to interactions with each other and with conditions, which are shaped by the consequences for the group.

Consider the evolution of the collective behavior that allows a school of fish to stay together as it moves and turns. Each fish responds to interaction with its nearby neighbor in the school. Natural selection acts on variation among individual fish in how they respond to each other and to the surroundings. Presumably, sometime in the evolutionary past, fish that were more attentive to the movement of neighbors, or who responded in a certain way, were more likely to reproduce than less attentive or less responsive fish. Natural selection might favor the attentive behavior if being in a school of fish that collectively turned faster made the fish more likely to escape predators.

To learn whether natural selection is currently still shaping this collective behavior, we would look for variation among fish in how they interact when they turn in response to the presence of predators, and how this affects the school's capacity to avoid predators. Next, we would need to ask whether this variation is heritable, by seeing whether the responses of fish offspring resemble those of their parents. Finally, we would ask whether the variation among fish, in their responses to interactions, is associated with differences among schools in how many offspring their fish survive to produce. If we could track all this, we would be able to trace how natural selection shapes the collective response of schools of fish to predators.

Tracing the evolution of collective behavior in cells proceeds along the same lines. For example, cancer as a disease is the result of the evolution of cancer cells from normal ones. Cancer cells work collectively, with each other and with the rest of the body. We identify the disease of cancer when their collective behavior produces outcomes sufficiently different from those produced by their ancestors. Many types of cancer form tumors, and in some types, cells leave one tumor to metastasize to other locations. Natural selection acts on variation among cells in how they interact with others to select for those that form tumors, in the same way that it acts on variation among fish in response to their neighbors, to select for those that participate in turning schools. This does not mean that the cancer cells are each acting separately, for their own advantage, against the rest of the body in which they operate; instead, they are carrying out, in an altered way, the same process that their ancestors did, with a different outcome.[33]

Social insect colonies are different from fish and cells, because the colony, not the individual fish or cell, is what reproduces. Ant colonies, consisting of one or more reproductive females or queens plus sterile female workers, reproduce by making reproductives, daughter queens, and males, who leave the nest to mate with reproductives of the opposite sex from other colonies. Those daughter queens, once they mate, can found new colonies. A new colony is thus the offspring of at least two parent colonies, the one that produced the queen of the new colony and the one that produced the male she mated with. (Since in many ant species a queen mates with more than one male, there may be more than two parent colonies.)

Since colonies, not ants, are the reproductive individuals, to consider the evolution of collective behavior in an ant colony we have to look first at variation among colonies in how they regulate behavior in response to changing conditions. We can then ask whether that variation among colonies is associated with how many offspring colonies they have, and whether the behavior is inherited by offspring colonies from their parent colonies.

We did this to learn how natural selection is shaping the collective behavior of harvester ant colonies. Colonies vary in how they regulate foraging in response to water loss. Some colonies conserve water by reducing foraging on dry days, while other colonies forage even when it is dry. In a study done in 2010, we found that the colonies that reduced foraging during a severe drought were more likely to produce offspring colonies.[34]

Variation among colonies arises from variation in how individuals of each colony respond to interactions. Harvester ant colonies have in common the general process that regulates foraging: an outgoing forager does not leave the nest until it meets enough returning foragers with food. The differences among colonies are in workers' responses to those interactions, which are associated with differences in water loss and in dopamine neurophysiology.[35]

Differences among colonies in how they regulate foraging seem to be inherited from mother to daughter colony, and we are currently investigating this. Parent and offspring colonies are not close enough together for the ants to interact. If the collective regulation of foraging of daughter colonies resembles that of their mother colonies, this is likely to be related to variation among colonies in physiology, such as in the hydrocarbons that determine water loss and in the metabolism of dopamine. Ants that return from the previous foraging trip feeling a little more dehydrated, and ants with less dopamine acting in their brains, seem to require more interactions from returning foragers before they leave the nest. These colonies are then more likely to reduce their foraging activity in dry conditions.

Thus, differences among colonies in the response of individual ants to interactions with other ants produce variation among colonies in how they regulate foraging so as to respond collectively to current food availability and humidity.

Over the timescale of years and even decades, how a harvester ant colony regulates its foraging activity has ecological consequences for its relation with its neighborhood of colonies.[36] A colony lives for twenty to thirty years, almost always in the same place, so the spatial configuration of a colony's neighborhood sets how much area it can use to forage. Colonies compete with their neighbors of the same species for foraging area. Sometimes the trails of neighboring colonies meet, because the foragers of both colonies are searching the same area for food. Whatever food one colony gets, the other loses. A new colony is unlikely to survive in a crowded neighborhood, and older ones are more likely to die when their foraging area is small.

The relations among neighboring harvester ant colonies are influenced by long-term changes in climate. When rainfall is high, food is more abundant, because the plants whose seeds the ants eat produce more flowers, yielding more fruits and seeds for the ants. As the drought has deepened in the US Southwest, harvester ant colonies need more foraging area to survive. If the drought continues, how colonies regulate foraging in dry conditions may become even more important. Colonies may no longer be able to afford to sacrifice food intake to conserve water, especially colonies with neighbors that forage even when it is dry. This will take decades to play out, because that is the timescale of colony reproduction and life span.

The Research Program

This book has outlined an approach to investigating collective behavior. The first step is to learn how interactions among individuals adjust outcomes to changing conditions. The goal is to explain, not how things are, but how they change; in nature, everything flows. An ecological perspective provides hypotheses, such as the ones outlined here, about how collective behavior works in relation with its surroundings. How quickly interactions adjust corresponds to gradients in stability and in the distribution of resources and demands. The feedback regime that stimulates and inhibits activity corresponds to gradients in stability and the energy flow required to operate in the surroundings. The modularity of the interaction network reflects gradients in stability and in the distribution of resources and demands.

Observing collective behavior as it normally happens, and learning about how it adjusts to ordinary fluctuations in conditions, makes it possible to identify how the participants interact and how their interactions respond to changing situations. This investigation requires either experiments that perturb the system or, in a version of a natural experiment, a comparative approach that contrasts systems whose surroundings change in different ways.

To understand collective behavior, there is no need to struggle to put back together what was never apart: inside and outside, organism and environment, or the interests of the individual and the interests of the collective. The whole appears to be more than the sum of the parts because the parts do not sum—they intertwine, jostle, and respond.

ACKNOWLEDGMENTS

My PhD adviser, John Gregg, and my postdoctoral mentor, Richard Lewontin, introduced me to the ideas that led to this book. Since then, my thinking about collective behavior has developed through talking and working with so many people that it would not be possible to list them all here. I am indebted to everyone that reached across a disciplinary boundary to explain their work to me, and to the Harvard Society of Fellows, the Santa Fe Institute, and the Center for Advanced Studies in Behavior at Stanford, for offering opportunities for those conversations. For recent discussions I'd especially like to thank George Bassel, Andrew Ewald, Jim Ferrell, Zev Gartner, Steve Haber, Sam Kortum, the late Bruno Latour, Naomi Leonard, Margaret Levi, Saket Navlakha, Josh Ober, Lucy O'Brien, Debra Satz, Sonia Sultan, and Mary Jane West-Eberhard. My understanding of dynamical systems relies on Fred Adler's patient and beautifully intuitive explanations; all errors are mine. I am grateful to Eric von Hippel for challenging me to generalize from my work on ants. Conversations with Donna Haraway and LeAnn Howard cut a path forward. Adrienne Mayor helped at every step, as a beacon and a guide. Thanks to my editor at Princeton University Press, Alison Kalett, who always offers excellent advice. The manuscript has been greatly improved by the comments of Henry Cerbone, Elisabeth Lloyd, Helen Longino, Thomas Pradeu, Elizabeth Whalley, Bob White, and anonymous reviewers. Finally, my heartfelt thanks to everyone in my multispecies family, who each provided different and essential forms of support.

NOTES

Chapter 1. Introduction

1. Donna J. Haraway, *Crystals, Fabrics, and Fields: Metaphors That Shape Embryos* (Berkeley, CA: North Atlantic Books, 2004).

2. Vanessa Barone and Deirdre C. Lyons, "Live Imaging of Echinoderm Embryos to Illuminate Evo-Devo," *Frontiers in Cell and Developmental Biology* 10 (2022): 1007775, DOI: 10.3389/fcell.2022.1007775.

3. Daniel J. Nicholson and John Dupré, eds. *Everything Flows: Towards a Processual Philosophy of Biology* (Oxford: Oxford University Press, 2018).

4. Denis M. Walsh, "The Affordance Landscape: The Spatial Metaphors of Evolution," in *Entangled Life: Organism and Environment in the Biological and Social Sciences*, edited by Gillian Barker, Eric Desjardins, and Trevor Pearce, 213–36 (Dordrecht: Springer, 2014).

5. Thanks to Tadashi Ikegami for an interesting discussion on this point.

6. Kevin J. Cheung and Andrew J. Ewald, "Illuminating Breast Cancer Invasion: Diverse Roles for Cell-Cell Interaction," *Current Opinion in Cell Biology* 30 (2014): 99–111.

7. Melinda Belisle, Kabir G. Peay, and Tadashi Fukami, "Flowers as Islands: Spatial Distribution of Nectar-Inhabiting Microfungi among Plants of *Mimulus aurantiacus*, a Hummingbird-Pollinated Shrub," *Microbial Ecology* 63 (2012): 711–18, https://doi.org/10.1007/s00248-011-9975-8.

8. Richard Levins, *Evolution in Changing Environments: Some Theoretical Explorations* (Princeton, NJ: Princeton University Press, 1968).

Chapter 2. The Ecology of Collective Behavior in Ants

1. Deborah M. Gordon, "The Ecology of Collective Behavior in Ants," *Annual Review of Entomology* 64 (2019): 35–50, https://doi.org/10.1146/annurev-ento-011118-111923.

2. Deborah M. Gordon, *Ant Encounters: Interaction Networks and Colony Behavior* (Princeton, NJ: Princeton University Press, 2010).

3. Michael J. Greene and Deborah M. Gordon, "Cuticular Hydrocarbons Inform Task Decisions," *Nature* 423, no. 32 (2003): 32.

4. Mekala Sundaram, Erik Steiner, and Deborah M. Gordon, "Rainfall, Neighbors, and Foraging: The Dynamics of a Population of Harvester Ant Colonies 1988–2019," *Ecological Monographs* 92, no. 2 (2022): e1503, https://doi.org/10.1002/ecm.1503.

5. Thanks to Thomas Pradeu for this analogy.

6. Deborah M. Gordon, "Dynamics of Task Switching in Harvester Ants," *Animal Behaviour* 38 (1989): 194–204.

7. Arjun Chandrasekhar, James A. R. Marshall, Cortnea Austin, Savet Navlakha, and Deborah M. Gordon, "Better Tired than Lost: Turtle Ant Trail Networks Favor Coherence over Shortest Paths," *PLoS Computational Biology* 17, no. 10 (2021): e1009523, https://doi.org/10.1371/journal.pcbi.1009523; Shivam Garg, Kirankumar Shiragur, Deborah M. Gordon, and Moses Charikar, "Distributed Algorithms from Arboreal Ants for the Shortest Path Problem," *Proceedings of the National Academy of Sciences* 120, no. 6 (2023): e2207959120v.

8. Deborah M. Gordon, "Local Regulation of Trail Networks of the Arboreal Turtle Ant, *Cephalotes goniodontus*," *American Naturalist* 190, no. 6 (2017), http://orcid.org/0000-0002-1090-9539.

Chapter 3. Collective Behavior

1. Henry James, *The Tragic Muse* (London: Macmillan and Co., 1909).

2. Helen E. Longino, "Interaction: A Case for Ontological Pluralism," *Interdisciplinary Science Reviews* 45, no. 3 (2020): 432–45, https://doi.org/10.1080/03080188.2020.1794385.

3. Ross Sager and Jung-Youn Lee, "Plasmodesmata in Integrated Cell Signaling: Insights from Development and Environmental Signals and Stresses," *Journal of Experimental Botany* 65 (2014): 6337–58, https://doi.org/10.1093/jxb/eru365.

4. Julian S. Huxley, "A Natural Experiment on the Territorial Instinct," *British Birds* 27 (1933): 270–77.

5. Damien R. Farine, Colin J. Garroway, and Ben C. Sheldon, "Social Network Analysis of Mixed-Species Flocks: Exploring the Structure and Evolution of Interspecific Social Behaviour," *Animal Behaviour* 84, no. 5 (2012): 1271–77.

6. Lynn Margulis, *Symbiosis in Cell Evolution: Life and Its Environment on the Early Earth* (San Francisco: W. H. Freeman and Co., 1981).

7. Thomas C. G. Bosch and Michael G. Hadfield, eds., *Cellular Dialogues in the Holobiont* (Abingdon, UK: Taylor and Francis Group/CRC Press, 2021).

8. Ron Sender, Shai Fuchs, and Ron Milo, "Are We Really Vastly Outnumbered? Revisiting the Ratio of Bacterial to Host Cells in Humans," *Cell* 164, no. 3 (2016): 337–40, https://doi.org/10.1016/j.cell.2016.01.013; Margaret McFall-Ngai, Michael G. Hadfield, Thomas C. G. Bosch, and Jennifer J. Wernegreen, "Animals in a Bacterial World: A New Imperative for the Life Sciences," *Proceedings of the National Academy of Science* 110, no. 9 (2013): 3229–36, https://doi.org/10.1073/pnas.1218525110.

9. Scott F. Gilbert, Jan Sapp and Alfred I. Tauber, "A Symbiotic View of Life: We Have Never Been Individuals," *Quarterly Review of Biology* 87, no. 4 (2012), https://doi.org/10.1086/668166.

10. Marnie Freckleton and Brian T. Nedved, "When Does Symbiosis Begin? Bacterial Cues Necessary for Metamorphosis in the Marine Polychaete *Hydroides elegans*," in Bosch and Hadfield, *Cellular Dialogues in the Holobiont*, 1–16.

11. Tamar L. Goulet, Clayton B. Cook, and Denis Goulet, "Effect of Short-Term Exposure to Elevated Temperatures and Light Levels on Photosynthesis of Different Host-Symbiont

Combinations in the *Aiptasia pallidal–Symbiodinium* Symbiosis," *Limnology and Oceanography* 50, no. 5 (2005): 1490–98, https://doi.org/10.4319/lo.2005.50.5.1490.

12. "Crow Helps Hedgehog to Cross the Street," posted by ViralHog, May 21, 2020, https://www.youtube.com/watch?v=FIqFiQ2MwfA.

13. "Fearless Honey Badger Takes on 6 Lions," posted by Lion Mountain TV, November 13, 2015, https://www.youtube.com/watch?v=NvlalDNxccw.

14. "Python, Honey Badger & Jackal Fight Each Other," posted by Caters Video, December 9, 2019, https://www.youtube.com/watch?v=JgKN3BuvC3E.

15. "Rare—A Badger and a Coyote Hunting Together," posted by Epic Nature Judy, July 29, 2015, https://www.youtube.com/watch?v=XVO4XIxjIEQ.

16. Anita Williams Woolley, Christopher F. Chabris, Alex Pentland, Nada Hasmi, and Thomas W. Malone, "Evidence for a Collective Intelligence Factor in the Performance of Human Groups," *Science* 33, no. 6004 (2010): 686–88, DOI: 10.1126/science.1193147.

17. Carl R. Shapiro, Genevieve M. Starke, and Dennice F. Gayme, "Turbulence and Control of Wind Farms," *Annual Review of Control, Robotics, and Autonomous Systems* 5, no. 1 (2022): 579–602, https://doi.org/10.1146/annurev-control-070221-114032; Karthik Gopalakrishnan and Hamsa Balakrishnan, "Control and Optimization of Air Traffic Networks," *Annual Review of Control, Robotics, and Autonomous Systems* 4, no. 1 (2021): 397–424, https://doi.org/10.1146/annurev-control-070720-080844.

18. Naomi Ehrich Leonard and Simon A. Levin, "Collective Intelligence as a Public Good," *Collective Intelligence* 1, no. 1 (2022), https://doi.org/10.1177/26339137221083293.

19. Hisashi Murakami, Claudio Feliciani, Yuta Nishiyama, and Katsuhiro Nishinari, "Mutual Anticipation Can Contribute to Self-Organization in Human Crowds," *Science Advances* 7, no. 12 (2021), DOI: 10.1126/sciadv.abe7758.

20. Erving Goffman, *Behavior in Public Places* (New York: Free Press, 1963).

21. Jürg Lamprecht, "Variable Leadership in Bar-Headed Geese (*Anser indicus*): An Analysis of Pair and Family Departures," *Behaviour* 122, nos. 1/2 (1992): 105–19, https://doi.org/10.1163/156853992X00336.

22. Yael Katz, Kolbjørn Tunstrøm, Christos C. Ioannou, Cristián Huepe, and Iain D. Couzin, "Inferring the Structure and Dynamics of Interactions in Schooling Fish," *Proceedings of the National Academy of Science* 108, no. 146 (2011): 18720–25, https://doi.org/10.1073/pnas.1107583108.

23. Adam Shellard and Roberto Mayor, "Rules of Collective Migration: From the Wildebeest to the Neural Crest," *Philosophical Transactions of the Royal Society B: Biological Sciences* 375, no. 1807 (2020), https://doi.org/10.1098/rstb.2019.0387.

24. Adam Shellard, András Szabo, Xavier Trepat, and Roberto Mayor, "Supracellular Contraction at the Rear of Neural Crest Cell Groups Drives Collective Chemotaxis," *Science* 362, no. 6412 (2018): 339–43. DOI: 10.1126/science.aau3301.

25. Mauricio Cantor, Paulo C. Simões-Lopes Fabio, and Fábio G. Daura-Jorge, "Spatial Consequences for Dolphins Specialized in Foraging with Fishermen," *Animal Behaviour* 139 (2018): 19–27, https://doi.org/10.1016/j.anbehav.2018.03.002.

26. Julieta A. Rosell, Mark E. Olson, Rebeca Aguirre-Hernández, and Francisco J. Sánchez-Sesma, "Ontogenetic Modulation of Branch Size, Shape, and Biomechanics Produces

Diversity across Habitats in the *Bursera simaruba* Clade of Tropical Trees," *Evolution and Development* 14, no. 5 (2012), 437–49, https://doi.org/10.1111/j.1525-142X.2012.00564.x.

27. Peter Stoll, Jacob Weiner, Helene Muller-Landau, Elke Müller, and Toshihiko Hara, "Size Symmetry of Competition Alters Biomass-Density Relationships," *Proceedings of the Royal Society B: Biological Sciences* 269, no. 1506 (2002), https://doi.org/10.1098/rspb.2002.2137.

28. Michiel Paul Veldhuis, Emilian Samwel Kihwele, Joris P.G.M. Cromsigt, Joseph O. Ogutu, J. Grant C. Hopcraft, N. Owen-Smith, and Han Olff, "Large Herbivore Assemblages in a Changing Climate: Incorporating Water Dependence and Thermoregulation," *Ecology Letters* 22, no. 10 (2019): 1536–46, https://doi.org/10.1111/ele.13350.

29. Douglas A. Bastos, Jansen Zuanon, Lúcia Rapp Py-Daniel, and Carlos David de Santana, "Social Predation in Electric Eels," *Ecology Evolution* 11, no. 3 (2021): 1088–1092, https://doi.org/10.1002/ece3.7121.

30. Kelly J. Benoit-Bird and William F. Gilly, "Coordinated Nocturnal Behavior of Foraging Jumbo Squid *Dosidicus gigas*," *Marine Ecology Progress Series* 455 (2012): 211–28, https://doi.org/10.3354/meps09664.

31. Deborah M. Gordon, "From Division of Labor to Collective Behavior," *Behavioral Ecology and Sociobiology* 70, no. 7 (2015): 1101–8, DOI: 10.1007/s00265-015-2045-3.

32. Michael J. F. Barresi and Scott F. Gilbert, *Developmental Biology*, 12th ed. (Oxford: Oxford University Press, 2019).

33. Alejandro Sánchez Alvarado and Shinya Yamanaka, "Rethinking Differentiation: Stem Cells, Regeneration, and Plasticity," *Cell* 157, no. 1 (2014): 110–19, https://doi.org/10.1016/j.cell.2014.02.041.

34. Brian C. R. Bertram, "Vigilance and Group Size in Ostriches," *Animal Behaviour* 28, no. 1 (1980): 278–86, https://doi.org/10.1016/S0003-3472(80)80030-3.

35. Deborah M. Gordon, *Ant Encounters: Interaction Networks and Colony Behavior* (Princeton, NJ: Princeton University Press, 2010).

36. Gordon, "The Ecology of Collective Behavior in Ants."

37. Deborah M. Gordon and Natasha J. Mehdiabadi, "Encounter Rate and Task Allocation in Harvester Ants," *Behavioral Ecology and Sociobiology* 45, no. 5 (1999): 370–77, https://www.jstor.org/stable/4601616.

38. Gordon, "Dynamics of Task Switching in Harvester Ants."

39. Zachary Yong Huang, Erika Plettner, and Gene E. Robinson, "Effects of Social Environment and Worker Mandibular Glands on Endocrine-Mediated Development in Honey Bees," *Journal of Comparative Physiology A* 183, no. 2 (1998): 143–52, DOI: 10.1007/s003590050242.

40. Marie P. Suver, Akira Mamiya, and Michael H. Dickinson, "Octopamine Neurons Mediate Flight-Induced Modulation of Visual Processing in *Drosophila*," *Current Biology* 22, no. 24 (2012): 2294–2302.

41. Michael L. Anderson, *After Phrenology: Neural Reuse and the Interactive Brain* (Cambridge, MA: MIT Press, 2014).

42. See, for example, Richard F. Betzel, John D. Medaglia, and Danielle S. Bassett, "Diversity of Meso-Scale Architecture in Human and Non-Human Connectomes," *Nature Communication* 9 (2018): article 346, https://doi.org/10.1038/s41467-017-02681-z.

Chapter 4. Beyond Emergence

1. Carl G. Hempel and Paul Oppenheim, "Studies in the Logic of Explanation," *Philosophy of Science* 15, no. 2 (1948): 135–75, https://doi.org/10.1086/286983.

2. Denis M. Walsh, "Objectcy and Agency: Toward a Methodological Vitalism," in *Everything Flows: Towards a Processual Philosophy of Biology*, edited by Daniel J. Nicholson and John Dupré (Oxford: Oxford University Press, 2018).

3. Rémy Chauvin, "Les lois de l'ergonomie chez les fourmis au cours du transport d'objects," *Comptes rendus de l'Académie des Sciences: Serie D: Sciences naturelles* 273 (1971): 1862–65.

4. Denis M. Walsh, *Organisms, Agency, and Evolution* (Cambridge: Cambridge University Press, 2016).

5. Isabelle Stengers, "The Cultivation of Ways of Overlapping: A Matter of Reclaiming," in *A Book of the Body Politic: Connecting Biology, Politics and Social Theory* (San Giorgio Dialogue 2017), edited by Bruno Latour, Simon Schaffer, and Pasquale Gagliardi (Venice: Fondazione Giorgio Cini, 2020).

6. Richard Levins, *Evolution in Changing Environments: Some Theoretical Explorations* (Princeton, N.J.: Princeton University Press, 1968).

7. Mark Granovetter, "Threshold Models of Collective Behavior," *American Journal of Sociology* 83, no. 6 (1978): 1420–43.

8. John J. Hopfield, "Neural Networks and Physical Systems with Emergent Collective Computational Abilities," *Proceedings of the National Academy of Science* 79, no. 8 (1982): 2554–58.

9. Jean-Louis Deneubourg, Serge Aron, S. Goss, Jacques M. Pasteels, and G. Duerinck, "Random Behaviour, Amplification Processes, and Number of Participants: How They Contribute to the Foraging Properties of Ants," *Physica* 22D, nos. 1–3 (1986): 176–86.

10. Shay Gueron, Simon A. Levin, and Daniel I. Rubenstein, "The Dynamics of Herds: From Individuals to Aggregations," *Journal of Theoretical Biology* 182, no. 1 (1996): 85–98, https://doi.org/10.1006/jtbi.1996.0144; Akira Okubo, "Dynamical Aspects of Animal Grouping: Swarms, Schools, Flocks, and Herds," *Advances in Biophysics* 22 (1986): 1–94, https://doi.org/10.1016/0065-227X(86)90003-1.

11. David J. T. Sumpter, *Collective Animal Behavior* (Princeton, NJ: Princeton University Press, 2010); Nancy A. Lynch, *Distributed Algorithms* (San Francisco: Morgan Kaufman Publishers, 1996).

12. Nicholas T. Ouellette and Deborah M. Gordon, "Goals and Limitations of Modeling Collective Behavior in Biological Systems," *Frontiers in Physics: Social Physics* (June 14, 2021), https://doi.org/10.3389/fphy.2021.687823.

13. Frederick R. Adler, *Modeling the Dynamics of Life: Calculus and Probability for Life Scientists*, 2nd ed. (Pacific Grove, CA: Brooks/Cole, 2005).

14. James E. Ferrell Jr., *Systems Biology of Cell Signalling: Recurring Themes and Quantitative Models* (Boca Raton, FL: CRC Press, 2022).

15. See, for example, Peter M. Kareiva and Nanoko Shigesada, "Analyzing Insect Movement as a Correlated Random Walk," *Oecologia* 56, nos. 2/3 (1983): 234–38, DOI: 10.1007/BF00379695.

16. Dan Gorbonos, Reuven Ianconescu, James G. Puckett, Rui Ni, Nicholas T. Ouellette, and Nir S. Gov, "Long-Range Acoustic Interactions in Insect Swarms: An Adaptive Gravity Model," *New Journal of Physics* 18, no. 7 (2016): 1–16, DOI: 10.1088/1367-2630/18/7/073042.

17. Naomi Ehrick Leonard, "Multi-Agent System Dynamics: Bifurcation and Behavior of Animal Groups," *Annual Reviews in Control* 38, no. 2 (2014): 171–83, DOI: 10.1016/j.arcontrol.2014.09.002.

18. Katz et al., "Inferring the Structure and Dynamics of Interactions in Schooling Fish."

19. Serina Chang, Emma Pierson, Pang Wei Koh, Jaline Gerardin, Beth Redbird, David Grusky, and Jure Leskovec, "Mobility Network Models of Covid-19 Explain Inequities and Inform Reopening," *Nature* 589, no. 7840 (2021): 82–87, https://doi.org/10.1038/s41586-020-2923-3.

20. Nicola Santoro, *Design and Analysis of Distributed Algorithms* (Hoboken, NJ: John Wiley & Sons, 2007).

21. Saket Navlakha and Ziv Bar-Joseph, "Distributed Information Processing in Biological and Computational Systems," *Communications of the Association for Computing Machinery* 58, no. 1 (2015): 94–102, https://doi.org/10.1145/2678280.

22. Wai-Leung Ng and Bonnie L. Bassler, "Bacterial Quorum-Sensing Network Architectures," *Annual Review of Genetics* 43 (2009): 197–222, DOI: 10.1146/annurev-genet-102108-134304.

23. See, for example, in development, John Gerhart and Marc Kirschner, *Cells, Embryos, and Evolution: Toward a Cellular and Developmental Understanding of Phenotypic Variation and Evolutionary Adaptability* (Oxford: Blackwell Science, 1997).

24. Reza Farhadifar, Charles F. Baer, Aurore-Cécile Valfort, Erik C. Andersen, Thomas Müeller-Reichert, Marie Delattre, and Daniel J. Needleman, "Scaling, Selection, and Evolutionary Dynamics of the Mitotic Spindle," *Current Biology* 25, no. 6 (2015): 723–40, DOI: 10.1016/j.cub.2014.12.060.

25. Richard N. McLaughlin Jr., Frank J. Poelwijk, Arjun Raman, Walraj S. Gosal, and Rama Ranganathan, "The Spatial Architecture of Protein Function and Adaptation," *Nature* 491 (2012): 138–42, https://doi.org/10.1038/nature11500.

26. Miriam Grace and Marc-Thorsten Hütt, "Regulation of Spatiotemporal Patterns by Biological Variability: General Principles and Applications to *Dictyostelium discoideum*." *PLoS Computational Biology* 11, no. (2015): e1004367, https://doi.org/10.1371/journal.pcbi.1004367.

27. Arjun Chandrasekhar, Deborah M. Gordon, and Saket Navlakha, "A Distributed Algorithm to Maintain and Repair the Trail Networks of Arboreal Ants," *Scientific Reports* 8 (2018): article 9297, https://doi.org/10.1038/s41598-018-27160-3.

28. Arjun Chandrasekhar, James A. R. Marshall, Cortnea Austin, Saket Navlakha, and Deborah M. Gordon, "Better Tired than Lost: Turtle Ant Trail Networks Favor Coherence over Shortest Paths," *PLoS Computational Biology* 17, no. 10 (2021): e1009523, DOI: 10.1371/journal.pcbi.1009523.

29. See, for example, Shelby J. Sturgis and Deborah M. Gordon, "Aggression Is Task-Dependent in the Red Harvester Ant (*Pogonomyrmex barbatus*)," *Behavioral Ecology* 24, no. 2 (2013): 532–39, https://doi.org/10.1093/beheco/ars194.

30. Fernando Esponda and Deborah M. Gordon, "Distributed Nestmate Recognition in Ants," *Proceedings of the Royal Society B: Biological Sciences* 282, no. 1806 (2015), https://doi.org/10.1098/rspb.2014.2838.

31. T'ai H. Roulston, Grzegorz Buczkowski, and J. Silverman, "Nestmate Discrimination in Ants: Effect of Bioassay on Aggressive Behavior," *Insectes Sociaux* 50, no. 2 (2003): 151–59, DOI: 10.1007/s00040-003-0624-1.

32. Sturgis and Gordon, "Aggression Is Task-Dependent in the Red Harvester Ant."

33. Deborah M. Gordon and Alan W. Kulig, "Founding, Foraging, and Fighting: Colony Size and the Spatial Distribution of Harvester Ant Nests," *Ecology* 77, no. 8 (1996): 2393–2409, https://doi.org/10.2307/2265741.

34. Melanie E. Moses, Judy L. Cannon, Deborah M. Gordon, and Stephanie Forrest, "Distributed Adaptive Search in T-Cells: Lessons from Ants," *Frontiers in Immunology* 10, no. 1357 (2019), https://doi.org/10.3389/fimmu.2019.01357.

35. Thomas Pradeu, *The Limits of the Self: Immunology and Biological Identity* (Oxford: Oxford University Press, 2012).

36. Hangjian Ling, Guillam E. McIvor, Kasper van der Vaart, Richard T. Vaughan, Alex Thornton, and Nicholas T. Ouellette, "Local Interactions and Their Group-Level Consequences in Flocking Jackdaws," *Proceedings of the Royal Society B: Biological Sciences* 286, no. 1906 (2019), https://doi.org/10.1098/rspb.2019.0865.

37. Gordon, "The Ecology of Collective Behavior in Ants."

38. Merlijn Staps and Corina E. Tarnita, "When Being Flexible Matters: Ecological Underpinnings for the Evolution of Collective Flexibility and Task Allocation," *Proceedings of the National Academy of Science* 119, no. 18 (2022): e2116066119, https://doi.org/10.1073/pnas.2116066119.

Chapter 5. Rate and Feedback

1. Arthur T. Winfree, *The Geometry of Biological Time* (New York: Springer, 2001).

2. Thomas D. Seeley and Craig A. Tovey, "Why Search Time to Find a Food-Storer Bee Accurately Indicates the Relative Rates of Nectar Collecting and Nectar Processing in Honey Bee Colonies," *Animal Behaviour* 47, no. 2 (1994): 311–16, https://doi.org/10.1006/anbe.1994.1044.

3. Michael A. Gil, Marissa L. Baskett, Stephan B. Munch, and Andrew M. Hein, "Fast Behavioral Feedbacks Make Ecosystems Sensitive to Pace and Not Just Magnitude of Anthropogenic Environmental Change," *Proceedings of the National Academy of Science* 117, no. 41 (2020): 25580–89, https://doi.org/10.1073/pnas.2003301117.

4. Xianrui Cheng and James E. Ferrell Jr., "Apoptosis Propagates through the Cytoplasm as Trigger Waves," *Science* 361, no. 6402 (2018): 607–12, DOI: 10.1126/science.aah4065.

5. Rick A. Relyea, "Fine-tuned Phenotypes: Tadpole Plasticity under 16 Combinations of Predators and Competitors," *Ecology* 85, no. 1 (2004): 172–79, https://doi.org/10.1890/03-0169.

6. Victoria E. Deneke and Stefano Di Talia, "Chemical Waves in Cell and Developmental Biology," *Journal of Cell Biology* 217, no. 4 (2018): 1193–1204, https://doi.org/10.1083/jcb

.201701158; Alessandro De Simone, Maya N. Evanitsky, Luke Hayden, Ben D. Cox, Julia Wang, Valerie A. Tornini, Jianhong Ou, Anna Chao, Kenneth D. Poss, and Stefano Di Talia, "Control of Osteoblast Regeneration by a Train of Erk Activity Waves," *Nature* 590, no. 7844 (2021): 129–33, DOI: 10.1038/s41586-020-03085-8.

7. Lucy M. Aplin, Damien R. Farine, Julie Morand-Ferron, Andrew Cockburn, Alex Thornton, and Ben C. Sheldon, "Experimentally Induced Innovations Lead to Persistent Culture via Conformity in Wild Birds," *Nature* 518, no. 7540 (2015): 538–41, DOI: 10.1038/nature13998.

8. Michael L. Kennedy, *The Jacobin Clubs in the French Revolution, 1793–1795* (New York: Berghan Books, 2000).

9. J. Philip Grime, *Plant Strategies and Vegetation Processes* (New York: John Wiley & Sons, 1979); J. Philip Grime, "Evidence for the Existence of Three Primary Strategies in Plants and Its Relevance to Ecological and Evolutionary Theory," *American Naturalist* 111, no. 982 (1977): 1169–94, https://www.jstor.org/stable/2460262.

10. Efrat Dener, Alex Kacelnik and Hagai Shemesh, "Pea Plants Show Risk Sensitivity," *Current Biology* 26, no. 13 (2016): 1763–67, https://doi.org/10.1016/j.cub.2016.05.008.

11. Martin S. Shapiro, Cynthia Schuck-Paim, and Alex Kacelnik, "Risk Sensitivity for Amounts of and Delay to Rewards: Adaptation for Uncertainty or By-product of Reward Rate Maximising?" *Behavioural Processes* 89, no. 2 (2012): 104–14, DOI: 10.1016/j.beproc.2011.08.016.

12. Uri Alon, *An Introduction to Systems Biology: Design Principles of Biological Circuits* (Abingdon, UK: Chapman & Hall/CRC Press, 2007).

13. Mikkel-Holger S. Sinding, et al., "Arctic-Adapted Dogs Emerged at the Pleistocene-Holocene Transition," *Science* 368, no. 6498 (2020): 1495–99, DOI: 10.1126/science.aaz8599.

14. Susanne Diekelmann and Jan Born, "The Memory Function of Sleep," *Nature Reviews Neuroscience* 11 (2010): 114–26, https://doi.org/10.1038/nrn2762.

15. Maksim V. Plikus, Julie Ann Mayer, Damon de la Cruz, Ruth E. Baker, Philip K. Maini, Robert Maxson, and Cheng-Ming Chuong, "Cyclic Dermal BMP Signalling Regulates Stem Cell Activation during Hair Regeneration," *Nature* 451, no. 7176 (2008): 340–44, DOI: 10.1038/nature06457.

16. Andrew Groover, "Gravitropisms and Reaction Woods of Forest Trees—Evolution, Function, and Mechanisms," *New Phytologist* 211, no. 3 (2016): 790–802, DOI: 10.1111/nph.13968.

17. Michael L. Anderson and Barbara L. Finlay, "Allocating Structure to Function: The Strong Links between Neuroplasticity and Natural Selection," *Frontiers in Human Neuroscience* 7 (2013): 918, DOI: 10.3389/fnhum.2013.00918.

18. Deborah M. Gordon, Rainer Rosengren, and Liselotte Sundstrom, "The Allocation of Foragers in Red Wood Ants," *Ecological Entomology* 17, no. 2 (1992): 114–20, https://doi.org/10.1111/j.1365-2311.1992.tb01167.x; Dominic D. R. Burns, Daniel W. Franks, Catherine Parr, and Elva J. H. Robinson, "Ant Colony Networks Adapt to Resource Disruption," *Journal of Animal Ecology* 92, no. 1 (2020): 143–52, https://doi.org/10.1111/1365-2656.13198.

19. Rainer Rosengren, *Foraging Strategy of Wood Ants (*Formica rufa *Group). 1. Age Polyethism and Topographic Traditions, Acta Zoologica Fennica,* vol. 149 (1977), 1–30 (Hensinki: Societas pro Fauna et Flora Fennica, 1977–).

20. John Davison, Timothy J. Roper, Charles J. Wilson, Matthew J. Heydon, and Richard J. Delahay, "Assessing Spatiotemporal Associations in the Occurrence of Badger-Human Conflict in England," *European Journal of Wildlife Research* 57, no. 1 (2011): 67–76, https://doi.org/10.1007/s10344-010-0400-2; John D. Reeve, Alain C. Frantz, Deborah A. Dawson, Terry Burke, and Timothy J. Roper, "Low Genetic Variability, Female-Biased Dispersal and High Movement Rates in an Urban Population of Eurasian Badgers *Meles meles*," *Journal of Animal Ecology* 77, no. 5 (2008): 905–15, DOI: 10.1111/j.1365-2656.2008.01415.x.

21. Noa Pinter-Wollman, Lynne A. Isbell, and Lynette A. Hart, "Assessing Translocation Outcome: Comparing Behavioral and Physiological Aspects of Translocated and Resident African Elephants (*Loxodonta africana*)," *Biological Conservation* 142, no. 5 (2009): 1116–24, https://doi.org/10.1016/j.biocon.2009.01.027.

22. Aparna Suvrathan, Hannah L. Payne, and Jennifer L. Raymond, "Timing Rules for Synaptic Plasticity Matched to Behavioral Function," *Neuron* 92, no. 5 (2016), 959–67, DOI: 10.1016/j.neuron.2016.10.022.

23. Michael J. Galko and Mark A. Krasnow, "Cellular and Genetic Analysis of Wound Healing in *Drosophila* Larvae," *PLoS Biology* 2, no. 8 (2004): e239, https://doi.org/10.1371/journal.pbio.0020239.

24. Alyssa J. Sargent, Derrick J. E. Groom, and Alejandro Rico-Guevara, "Locomotion and Energetics of Divergent Foraging Strategies in Hummingbirds: A Review," *Integrative and Comparative Biology* 61, no. 2 (2021): 736–48, https://doi.org/10.1093/icb/icab124.

25. See, for example, Thomas S. Diebenstock and Iain D. Couzin, "Collective Behavior in Cancer Cell Populations," *BioEssays* 31, no. 2 (2009):190–97, https://doi.org/10.1002/bies.200800084; Guillermina R. Ramirez–San Juan, Patrick W. Oakes, and Margaret L. Gardel, "Contact Guidance Requires Leading-Edge Control of Spatial Protrusion," *Molecular Biology of the Cell* 28, no. 8 (2017): 1043–53, DOI: 10.1091/mbc.E16-11-0769.

26. Thanks to Sam Crow for this example.

27. Jean-Louis Deneubourg, Sergei Aron, Simon Goss, Jacques M. Pasteels, and G. Duerinck, "Random Behaviour, Amplification Processes, and Number of Participants: How They Contribute to the Foraging Properties of Ants," *Physica D: Nonlinear Phenomena* 22, nos. 1–3 (1986): 176–86.

28. See, for example, Lauren Ancel Meyers, Michael E. J. Newman, Michael Martin, and Stephanie Schrag, "Applying Network Theory to Epidemics: Control Measures for *Mycoplasma pneumoniae* Outbreaks," *Emerging Infectious Diseases* 9, no. 2 (2003): 204–10.

29. John J. Tyson, Katherine C. Chen, and Bela Novak, "Sniffers, Buzzers, Toggles, and Blinkers: Dynamics of Regulatory and Signaling Pathways in the Cell," *Current Opinion in Cell Biology* 15, no. 2 (2003): 221–31, https://doi.org/10.1016/S0955-0674(03)00017-6.

30. Thanks to Lucy O'Brien for this example.

31. Max Gassmann, Heimo Mairbäurl, Leonid Livshits, Svenja Seide, Matthes Hackbusch, Monika Malczyk, Simone Kraut, Norina N. Gassmann, Norbert Weissmann, and

Martina U. Muckenthaler, "The Increase in Hemoglobin Concentration with Altitude Varies among Human Populations," *Annals of the New York Academy of Sciences* 1450, no. 1 (2019): 204–20, https://doi.org/10.1111/nyas.14136.

32. Richard FitzHugh, "Impulses and Physiological States in Theoretical Models of Nerve Membrane," *Biophysical Journal* 1, no. 6 (1961): 445, https://doi.org/10.1016/S0006-3495(61)86902-6; J. Nagumo, S. Arimoto, and S. Yoshizawa, "An Active Pulse Transmission Line Simulating Nerve Axon," *Proceedings of the IEEE* 50, no. 10 (1962): 2061–70, http://dx.doi.org/10.1109/JRPROC.1962.288235; James E. Ferrell Jr., *Systems Biology of Cell Signaling: Recurring Themes and Quantitative Models* (Boca Raton, FL: CRC Press, 2022).

33. Jacob D. Davidson, Roxana P. Auraco-Aliaga, Sam Crow, Deborah M. Gordon, and Mark S. Goldman, "Effect of Interactions between Harvester Ants on Forager Decisions," *Frontiers in Ecology and Evolution* (October 5, 2016), https://doi.org/10.3389/fevo.2016.00115.

34. Andreas Doncic and Jan M. Skotheim, "Feedforward Regulation Ensures Stability and Rapid Reversibility of a Cellular State," *Molecular Cell* 50, no. 6 (2013): 856–68, https://doi.org/10.1016/j.molcel.2013.04.014.

35. Jonathan L. Payne, John R. Grovers, Adam B. Jost, Thienan Nguyen, Sarah E. Moffitt, Tessa M. Hill, and Jan M. Skotheim, "Late Paleozoic Fusilinoidan Gigantism Driven by Atmospheric Hyperoxia," *Evolution* 66, no. 9 (2012): 2929–39, https://doi.org/10.1111/j.1558-5646.2012.01626.x.

36. Charles Whittaker and Caroline Dean, "The FLC Locus: A Platform for Discoveries in Epigenetics and Adaptation," *Annual Review of Cell and Developmental Biology* 33 (2017): 555–75, DOI: 10.1146/annurev-cellbio-100616-060546.

37. Jackson Liang, Shruthi Balachandra, Sang Ngo, Lucy Erin O'Brien, "Feedback Regulation of Steady-State Epithelial Turnover and Organ Size," *Nature* 548, no. 7669 (2017): 588–91, https://doi.org/10.1038/nature23678.

38. N. B. Davies and A. I. Houston, "Owners and Satellites: The Economics of Territory Defence in the Pied Wagtail, *Motacilla alba*," *Journal of Animal Ecology* 50, no. 1 (1981): 157–80, https://doi.org/10.2307/4038.

39. Enrique López-Juez, "Steering the Solar Panel: Plastids Influence Development," *New Phytologist* 182, no. 2 (2009): 287–90, DOI: 10.1111/j.1469-8137.2009.02808.x.

40. Nicole E. Heller and Deborah M. Gordon, "Seasonal Spatial Dynamics and Causes of Nest Movement in Colonies of the Invasive Argentine Ant (*Linepithema humile*)," *Ecological Entomology* 31, no. 5 (2006): 499–510, https://doi.org/10.1111/j.1365-2311.2006.00806.x.

41. Sarah E. Bengston and Anna Dornhaus, "Latitudinal Variation in Behaviors Linked to Risk Tolerance Is Driven by Nest-Site Competition and Spatial Distribution in the Ant *Temnothorax rugatulus*." *Behavioral Ecology Sociobiology* 69, no. 8 (2015): 1265–74, https://www.jstor.org/stable/43599631.

42. David E. Cade, Nicholas Carey, Paolo Domenici, Jean Potvin, and Jeremy A. Goldbogen, "Predator-Informed Looming Stimulus Experiments Reveal How Large Filter Feeding Whales Capture Highly Maneuverable Forage Fish," *Proceedings of the National Academy of Science* 117, no. 1 (2020): 472–78, https://doi.org/10.1073/pnas.1911099116.

43. Christian Rutz et al., "Covid-19 Lockdown Allows Researchers to Quantify the Effects of Human Activity on Wildlife," *Nature Ecology and Evolution* 4, no. 9 (2020): 1156–59, https://doi.org/10.1038/s41559-020-1237-z.

44. Marlee A. Tucker et al., "Moving in the Anthropocene: Global Reductions in Terrestrial Mammalian Movements," *Science* 359, no. 6374 (2018): 466–69, DOI: 10.1126/science.aam9712.

45. Domitilla Del Vecchio and Eduardo D. Sontag, "Synthetic Biology: A Systems Engineering Perspective," in *Control Theory and Systems Biology*, edited by Pablo A. Iglesias and Brian P. Ingalls (Cambridge, MA: MIT Press, 2010), 101–24.

46. Cynthia Jane Zabel and Spencer James Taggart, "Shift in Red Fox, *Vulpes*, Mating System Associated with El Niño in the Bering Sea," *Animal Behaviour* 38, no. 5 (1989): 830–38, https://doi.org/10.1016/S0003-3472(89)80114-9.

47. David W. Macdonald, Patrick Doncaster, Malcolm Newdick, Heribert Hofer, Fiona Mathews, and Paul J. Johnson, "Foxes in the Landscape," in *Wildlife Conservation on Farmland*, vol. 2: *Conflict in the Countryside*, edited by David W. Macdonald and Ruth E. Feber (Oxford: Oxford University Press, 2015).

48. Aplin et al., "Experimentally Induced Innovations Lead to Persistent Culture via Conformity in Wild Birds."

Chapter 6. Modularity

1. Mark E. J. Newman, *Networks: An Introduction* (Oxford: Oxford University Press, 2010).

2. Martin N. Muller and Richard W. Wrangham, "Dominance, Aggression, and Testosterone in Wild Chimpanzees: A Test of the 'Challenge Hypothesis,'" *Animal Behaviour* 67, no. 1 (2004): 113–23, https://doi.org/10.1016/j.anbehav.2003.03.013.

3. Herbert A. Simon, "Foreword," in *Modularity: Understanding the Development and Evolution of Natural Complex Systems*, edited by Werner Callebaut and Diego Rasskin-Gutman (Cambridge, MA: MIT Press, 2005), ix–xiv.

4. Emilie C. Snell-Rood, James David Van Dyken, Tami Cruikshank, Michael J. Wade, and Armin P. Moczek, "Toward a Population Genetic Framework of Developmental Evolution: The Costs, Limits and Consequences of Phenotypic Plasticity," *BioEssays* 32, no. 1 (2010): 71–81, https://doi.org/10.1002/bies.200900132.

5. Paulo R. Guimarães Jr., "The Structure of Ecological Networks across Levels of Organization," *Annual Review of Ecology, Evolution, and Systematics* 51 (2020): 433–60, https://doi.org/10.1146/annurev-ecolsys-012220-120819.

6. John Gerhart and Marc Kirschner, "The Theory of Facilitated Variation," *Proceedings of the National Academy of Science* 104, supp. 1 (2007): 8582–89, https://doi.org/10.1073/pnas.0701035104.

7. See, for example, Jens Krause, Richard James, Daniel W. Franks, and Darren P. Croft, eds., *Animal Social Networks* (Oxford: Oxford University Press, 2014); Sara Brin Rosenthal, Colin R. Twomey, Andrew T. Hartnett, Hai Shan Wu, and Iain D. Couzin, "Revealing the Hidden Networks of Interaction in Mobile Animal Groups Allows Prediction of Complex

Behavioral Contagion," *Proceedings of the National Academy of Science* 112, no. 15 (2015): 4690–95, https://doi.org/10.1073/pnas.1420068112.

8. Herbert A. Simon, *The Sciences of the Artificial* (Cambridge, MA: MIT Press, 1996).

9. See, for example, Günter P. Wagner, "Homologues, Natural Kinds, and the Evolution of Modularity," *American Zoologist* 36, no. 1 (1996): 36–43, https://doi.org/10.1093/icb/36.1.36.

10. Jeff Clune, Jean-Baptiste Mouret, and Hod Lipson, "The Evolutionary Origins of Modularity," *Proceedings of the Royal Society B: Biological Sciences* 280, no. 1755 (2013): 20122863, http://dx.doi.org/10.1098/rspb.2012.2863.

11. Daniel A. Levinthal and James G. March, "The Myopia of Leanring", *Strategic Management Journal* 14 (1993): 95–112, https://doi.org/10.1002/smj.4250141009.

12. See, for example, Callebaut and Rasskin-Gutman, *Modularity*; Hans de Kroon, Heidrun Huber, Josef F. Stuefer, and Jan M. van Groenendael, "A Modular Concept of Phenotypic Plasticity in Plants," *The New Phytologist* 166 (2005): 73–82; Mark E. Olson and Julieta A. Rosell, "Using Heterochrony to Infer Modularity in the Evolution of Stem Diversity in *Moringa* (Moringaceae)," *Evolution* 60, no. 4 (2006): 724–34, https://doi.org/10.1111/j.0014-3820.2006.tb01151.x.

13. Nadav Kashtan and Uri Alon, "Spontaneous Evolution of Modularity and Network Motifs," *Proceedings of the National Academy of Science* 102, no. 39 (2005): 13773–78, https://doi.org/10.1073/pnas.0503610102.

14. See, for example, Mary Jane West-Eberhard, *Developmental Plasticity and Evolution* (Oxford: Oxford University Press, 2003); Günter P. Wagner, Mihaela Pavličev, and James M. Cheverud, "The Road to Modularity," *Nature Reviews: Genetics* 8 (2007): 921–31, DOI: 10.1038/nrg2267.

15. Roberto Salguero-Gómez and Brenda B. Casper, "A Hydraulic Explanation for Size-Specific Plant Shrinkage: Developmental Hydraulic Sectoriality," *New Phytologist* 89, no. 1 (2011): 229–40, DOI: 10.1111/j.1469-8137.2010.03447.x.

16. Tatiana P. Flanagan, Noa M. Pinter-Wollman, Melanie E. Moses, and Deborah M. Gordon, "Fast and Flexible: Argentine Ants Recruit from Nearby Trails," *PLoS One* 8, no. 8 (2013): e70888, https://doi.org/10.1371/journal.pone.0070888.

17. Melanie E. Moses, Judy L. Cannon, Deborah M. Gordon and Stephanie Forrest, "Distributed Adaptive Search in T-Cells: Lessons from Ants," *Frontiers in Immunology* 10, no. 1357 (2019), https://doi.org/10.3389/fimmu.2019.01357.

18. Danielle S. Bassett and Edward T. Bullmore, "Small-World Brain Networks," *The Neuroscientist* 23, no. 6 (2017): 499–516, DOI: 10.1177/1073858406293182; Urs Braun, Anais Harneit, Giulio Pergola, et al., "Brain Network Dynamics during Working Memory Are Modulated by Dopamine and Diminished in Schizophrenia," *Nature Communications* 12, no. 3478 (2021), https://doi.org/10.1038/s41467-021-23694-9.

19. Bruce Wang, Ludan Zhao, Matt Fish, Catriona Y. Logan, and Roel Nusse, "Self-Renewing Diploid Axin2(+) Cells Fuel Homeostatic Renewal of the Liver," *Nature* 524, no. 7564 (2015), DOI: 10.1038/nature14863.

20. Andrew Groover, "Gravitropisms and Reaction Woods of Forest Trees—Evolution, Function, and Mechanisms," *New Phytologist* 211, no. 3 (2016): 790–802, DOI: 10.1111/nph.13968.

21. Daniel Lobo, Wendy S. Beane, and Michael Levin, "Modeling Planarian Regeneration: A Primer for Reverse-Engineering the Worm," *PLoS Computational Biology* 8, no. 4 (2012), DOI: 10.1371/journal.pcbi.1002481.

22. Robert A. Hinde and J Stevenson-Hinde, "Towards Understanding Relationships: Dynamic Stability," in *Growing Points in Ethology*, edited by P.P.G. Bateson and R. A. Hinde (Cambridge: Cambridge University Press, 1976).

23. Ivan D. Chase, Costanza Bartolomeo, and Lee A. Dugatkin, "Aggressive Interactions and Inter-Contest Interval: How Long Do Winners Keep Winning?" *Animal Behaviour* 48, no. 2 (1994): 393–400, https://doi.org/10.1006/anbe.1994.1253; W. Brent Lindquist and Ivan D. Chase, "Data-Based Analysis of Winner-Loser Models of Hierarchy Formation in Animals," *Bulletin of Mathematical Biology* 71, no. 3 (2009): 556–84, DOI: 10.1007/s11538-008-9371-9.

24. Darren .P Croft, Richard James, and Jens Krause, *Exploring Animal Social Networks* (Princeton, NJ: Princeton University Press, 2008); David Lusseau and M.E.J. Newman, "Identifying the Role That Animals Play in Their Social Networks," *Proceedings of the Royal Society B: Biological Sciences* 271, supp. 6 (2004): S477–81, http://doi.org/10.1098/rsbl.2004.0225.

25. Mark S. Granovetter, "The Strength of Weak Ties," *American Journal of Sociology* 78, no. 6 (1973): 1360–80.

26. Thanks to Jason Watters, Vice President for Wellness and Animal Behavior, San Francisco Zoological Society.

27. T. H. Clutton-Brock and Paul H. Harvey, "Primate Ecology and Social Organization," *Journal of Zoology* 183, no. 1 (1977): 1–39, https://doi.org/10.1111/j.1469-7998.1977.tb04171.x.

28. A. Catherine Markham, Vishwesha Guttal, Susan C. Alberts, and Jeanne Altmann, "When Good Neighbors Don't Need Fences: Temporal Landscape Partitioning among Baboon Social Groups," *Behavioral Ecology and Sociobiology* 67 (2013): 875–84, https://doi.org/10.1007/s00265-013-1510-0.

29. Gabriel Ramos-Fernandez, Sandra E. Smith Aguilar, David C. Krakauer, and Jessica C. Flack, "Collective Computation in Animal Fission-Fusion Dynamics," *Frontiers in Robotics and AI* 7, no. 90 (2020), https://doi.org/10.3389/frobt.2020.00090.

30. Charles Janson, Maria Celia Baldovino, and Mario Di Bitetti, "The Group Life Cycle and Demography of Brown Capuchin Monkeys (*Cebus [apella] nigritus*) in Iguazú National Park, Argentina," In *Long-Term Field Studies of Primates*, edited by Peter M. Kappeler and David P. Watts (Berlin: Springer-Verlag, 2012).

31. Edward N. Luttwak, *The Grand Strategy of the Roman Empire: From the First Century CE to the Third* (Baltimore: Johns Hopkins University Press, 2016).

32. Stephen Haber, Roy Elis, and Jordan Horrillo, "The Ecological Origins of Economic and Political Systems," November 17, 2021, https://papers.ssrn.com/sol3/papers.cfm?abstract_id=3958073.

33. See, for example, Jeremy Freeman, Nikita Vladimirov, Takashi Kawashima, Yu Mu, Nicholas J. Sofroniew, Davis V. Bennett, Joshua Rosen, Chao-Tsung Yang, Loren L. Looger, and Misha B. Ahrens, "Mapping Brain Activity at Scale with Cluster Computing," *Nature Methods* 11 (2014): 941–50, https://doi.org/10.1038/nmeth.3041.

34. Armin P. Moczek, Sonia Sultan, Susan Foster, Cris Ledón-Rettig, Ian Dworkin, H. Fred Nijhout, Ehab Abouheif, and David W. Pfennig, "The Role of Developmental Plasticity in Evolutionary Innovation," *Proceedings of the Royal Society B: Biological Sciences* 2011, no. 278 (2011): 2705–13, https://doi.org/10.1098/rspb.2011.0971.

35. Daizaburo Shizuka, Alexis S. Chaine, Jennifer Anderson, Oscar Johnson, Inger Marie Laursen, and Bruce E. Lyon, "Across-Year Social Stability Shapes Network Structure in Wintering Migrant Sparrows," *Ecology Letters* 17, no. 8 (2014): 998–1007, DOI: 10.1111/ele.12304.

36. Siva R. Sundaresan, Ilya R. Fischoff, Jonathan Dushoff, and Daniel I. Rubenstein, "Network Metrics Reveal Differences in Social Organization between Two Fission-Fusion Species, Grevy's Zebra and Onager," *Oecologia* 151, no. 1 (2007): 140–49, DOI: 10.1007/s00442-006-0553-6.

37. Zuleyma Tang-Martinez and T. Brandt Ryder, "The Problem with Paradigms: Bateman's Worldview as a Case Study," *Integrative and Comparative Biology* 45, no. 5 (2006): 821–30, https://www.jstor.org/stable/4485865.

38. The series includes: Torgrim Breiehagen and Tore Slagsvold, "Male Polyterritoriality and Female-Female Aggression in Pied Flycatchers *Ficedula hypoleuca*," *Animal Behaviour* 36, no. 2 (1988): 604–6, https://doi.org/10.1016/S0003-3472(88)80033-2; and Glenn-Peter Sætre, Svein Dale, and Tore Slagsvold, "Female Pied Flycatchers Prefer Brightly Coloured Males" *Animal Behaviour* 48, no. 6 (1994): 1407–16, https://doi.org/10.1006/anbe.1994.1376.

39. David Canal, Lotte Schlicht, Simone Santoro, Carlos Camacho, Jesús Martínez-Padilla, and Jaime Potti, "Phenology-Mediated Effects of Phenotype on the Probability of Social Polygyny and Its Fitness Consequences in a Migratory Passerine," *BMC Ecology and Evolution* 21, no. 1 (2021): article 55, https://doi.org/10.1186/s12862-021-01786-w.

40. Arne Lundberg and Rauno V. Alatalo, *The Pied Flycatcher* (London: T & AD Poyser Ltd., 2010).

Chapter 7. Investigating How Collective Behavior Works

1. Frederick R. Adler and Deborah M. Gordon, "Cancer Ecology and Evolution: Positive Interactions and System Vulnerability," *Current Opinion in Systems Biology* 17 (2019): 1–7.

2. Charles Darwin, *The Formation of Vegetable Mould through the Action of Worms: With Observations on Their Habits* (London: John Murray, 1881), chap. 1, 12–14.

3. Elena Rondeau, Nicolas Larmonier, Thomas Pradeu, and Andreas Bikfalvi, "Characterizing Causality in Cancer," *eLife* 8 (2019): e53755, https://doi.org/10.7554/eLife.53755.

4. Tadashi Fukami, "Historical Contingency in Community Assembly: Integrating Niches, Species Pools, and Priority Effects," *Annual Review of Ecology, Evolution, and Systematics* 46 (2015): 1–23, https://doi.org/10.1146/annurev-ecolsys-110411-160340.

5. Kimberle Crenshaw, "Demarginalizing the Intersection of Race and Sex: A Black Feminist Critique of Antidiscrimination Doctrine, Feminist Theory, and Antiracist Politics," *University of Chicago Legal Forum* 140 (1989): 25–42, DOI: 10.4324/9781315582924-10.

6. Paul B. Rainey, Nicolas Desprat, William W. Driscoll, and Xue-Xian Zhang, "Microbes Are Not Bound by Sociobiology: Response to Kümmerli and Ross-Gillespie," *Evolution* 68, no. 11 (2013): 3344–55, 3345, DOI: 10.1111/evo.12508.

7. Mam Y. Mboge and Mina J. Bissell, "The Not-So-Sweet Side of Sugar: Influence of the Microenvironment on the Processes That Unleash Cancer," *Biochimica et Biophysica Acta: Molecular Basis of Disease* 1866, no. 12 (2020): 165960, https://doi.org/10.1016/j.bbadis.2020.165960.

8. See, for example, Kevin J. Cheung, Edward Gabrielson, Zena Werb, and Andrew J. Ewald, "Collective Invasion in Breast Cancer Requires a Conserved Basal Epithelial Program," *Cell* 155, no. 7 (2013): 1639–51, DOI: 10.1016/j.cell.2013.11.029.

9. Deborah M. Gordon, "The Dynamics of the Daily Round of the Harvester Ant Colony (*Pogonomyrmex barbatus*)," *Animal Behaviour* 34, no. 5 (1986): 1402–19, https://doi.org/10.1016/S0003-3472(86)80211-1.

10. Gordon, "Dynamics of Task Switching in Harvester Ants."

11. Deborah M. Gordon, "Behavioral Flexibility and the Foraging Ecology of Seed-Eating Ants," *American Naturalist* 138, no. 2 (1991): 379–411, DOI: 10.1086/285223.

12. Deborah M. Gordon, "The Regulation of Foraging Activity in Red Harvester Ant Colonies," *American Naturalist* 159, no. 5 (2002): 509–18, https://doi.org/10.1086/339461.

13. Robert J. Schafer, Susan Holmes, and Deborah M. Gordon, "Forager Activation and Food Availability in Harvester Ants," *Animal Behaviour* 71 (2006): 815–22, DOI: 10.1016/j.anbehav.2005.05.024.

14. Balaji Prabhakar, Katherine N. Dektar, and Deborah M. Gordon, "The Regulation of Ant Colony Foraging Activity without Spatial Information," *PLoS Computational Biology* 8, no. 8 (2012): e1002670, https://doi.org/10.1371/journal.pcbi.1002670.

15. Michael J. Greene, Noa Pinter-Wollman, and Deborah M. Gordon, "Interactions with Combined Chemical Cues Inform Harvester Ant Foragers' Decisions to Leave the Nest in Search of Food," *PLoS ONE* 8, no. 1 (2013): e52219, https://doi.org/10.1371/journal.pone.0052219.

16. Daniel A. Friedman, Anna Pilko, Dorota Skowronska-Krawczyk, Karolina Krasinska, Jacqueline W. Parker, Jay Hirsh, and Deborah M. Gordon, "The Role of Dopamine in the Collective Regulation of Foraging in Harvester Ants," *iScience* 8 (2018): 283–94, DOI: 10.1016/j.isci.2018.09.001.

Chapter 8. Evolution, Heredity, and the Causes of Behavior

1. Deborah M. Gordon, "The Ecology of Collective Behavior," *PloS Biology* 12, no. 3 (2014): e1001805, https://doi.org/10.1371/journal.pbio.1001805; Deborah M. Gordon, "The Evolution of the Algorithms for Collective Behavior," *Cell Systems* 3, no. 6 (2016): P514–20, https://doi.org/10.1016/j.cels.2016.10.013.

2. Renée A. Duckworth, "The Role of Behavior in Evolution: A Search for Mechanism," *Evolutionary Ecology* 23, no 4 (2009): 513–31, https://doi.org/10.1007/s10682-008-9252-6.

3. See, for example, Moczek et al., "The Role of Developmental Plasticity in Evolutionary Innovation; Emilie C. Snell-Rood and Sean M. Ehlman, "Ecology and Evolution of Plasticity," in *Phenotypic Plasticity and Evolution*, edited by David W. Pfennig (Boca Raton, FL: CRC Press, 2021).

4. Richard C. Lewontin, *The Triple Helix: Gene, Organism, and Environment* (Cambridge, MA: Harvard University Press, 2000).

5. F. John Odling-Smee, Kevin N. Laland, and Marcus W. Feldman, *Niche Construction: The Neglected Process in Evolution* (Princeton, NJ: Princeton University Press, 2003); Kevin N. Laland, F. John Odling-Smee, and Marcus W. Feldman, "Evolutionary Consequences of Niche Construction and Their Implications for Ecology," *Proceedings of the National Academy of Science* 96, no. 18 (1999): 10242–47, DOI: 10.1073/pnas.96.18.10242.

6. See, for example, Sonia E. Sultan, *Organism and Environment: Ecological Development, Niche Construction, and Adaptation* (Oxford: Oxford University Press, 2015); Denis M. Walsh, "Situated Adaptationism," in *The Environment*, edited by William P. Kabasenche, Michael O'Rourke, and Matthew H. Slater (Cambridge, MA: MIT Press, 2012); Mark E. Olson, "The Developmental Renaissance in Adaptationism," *Trends in Ecology and Evolution* 27, no. 5 (2012): 278–87, https://doi.org/10.1016/j.tree.2011.12.005.

7. Sultan, *Organism and Environment*, 49.

8. Lewontin, *The Triple Helix*.

9. Günter Vogt, Martin Huber, Markus Thiemann, Gerald van den Boogaart, Oliver J. Schmitz, and Christof D. Schubart, "Production of Different Phenotypes from the Same Genotype in the Same Environment by Developmental Variation," *Journal of Experimental Biology* 211, pt. 4 (2008): 510–23, DOI: 10.1242/jeb.008755.

10. Carl D. Schlichting and Massimo Pigliucci, *Phenotypic Evolution: A Reaction Norm Perspective* (Sunderland, MA: Sinauer Press, 1998).

11. Alessandro Minelli, "Lichens and Galls: Two Families of Chimeras in the Space of Form," *Revista de Filosofia* 19 (2017): 91–105.

12. Sewall Wright, "Evolution in Mendelian Populations," *Genetics* 16, no. 2 (1931): 97–159, https://doi.org/10.1093/genetics/16.2.97; John N. Thompson, *Relentless Evolution* (Chicago: University of Chicago Press, 2013).

13. Daniel A. Levinthal and James G. March, "The Myopia of Learning," *Strategic Management Journal* 14, no. S2 (1993): 95–112, https://doi.org/10.1002/smj.4250141009.

14. See, for example, Snell-Rood et al., "Toward a Population Genetic Framework of Developmental Evolution."

15. West-Eberhard, *Developmental Plasticity and Evolution*.

16. Ibid., 4.

17. See, for example, Giuseppe Fusco, ed., *Perspectives on Evolutionary and Developmental Biology: Essays for Alessandro Minelli* (Padua, Italy: Padua University Press, 2019).

18. Gilbert Gottlieb, "Species Identification by Avian Neonates: Contributory Effect of Perinatal Auditory Stimulation," *Animal Behaviour* 14, nos. 2/3 (1966): 282–90, https://doi.org/10.1016/S0003-3472(66)80084-2.

19. Jack London, *The Call of the Wild* (New York: MacMillan, 1903; Dover, 1990).

20. Ibid., 14.

21. Marjorie Grene, *Descartes* (Minneapolis: University of Minnesota Press, 1985).

22. Bruno Latour, *We Have Never Been Modern* (Cambridge, MA: Harvard University Press, 1993).

23. Thanks to Kurt Smith for this insight.

24. Josiah Ober, *The Greeks and the Rational: The Discovery of Practical Reason* (Berkeley: University of California Press, 2022).

25. Lewontin, *The Triple Helix*, 48–49.

26. Walsh, "The Affordance Landscape."

27. James J. Gibson, *The Ecological Approach to Visual Perception* (Boston: Houghton Mifflin, 1979), 127.

28. Mina Bissell and William C. Hines, "Why Don't We Get More Cancer? A Proposed Role of the Microenvironment in Restraining Cancer Progression," *Nature Medicine* 17, no. 3 (2011): 320–29, https://doi.org/10.1038/nm.2328; Héctor Peinado, Haiying Zhang, Irina R. Matei, et al., "Pre-metastatic Niches: Organ-Specific Homes for Metastases," *Nature Reviews: Cancer* 17 (2017): 302–17, https://doi.org/10.1038/nrc.2017.6.

29. Iñigo Martincorena, Amit Roshan, Moritz Gerstung, Peter Ellis, Peter V. Loo, et al., "Tumor Evolution: High Burden and Pervasive Positive Selection of Somatic Mutations in Normal Human Skin," *Science* 348, no. 6237 (2015): 880–86, DOI: 10.1126/science.aaa6806, Alec E. Cerchiari, James C. Garbe, Noel Y. Jee, Michael E. Todhunter, Kyle E. Broaders, et al., "A Strategy for Tissue Self-Organization That Is Robust to Cellular Heterogeneity and Plasticity," *Proceedings of the National Academy of Science* 112, no. 7 (2015): 2287–92, https://doi.org/10.1073/pnas.1410776112.

30. Nikolaas Tinbergen, *The Herring-Gull's World: A Study of the Social Behavior of Birds* (London: Collins, 1953), 63–64.

31. Adrienne Mayor, *Gods and Robots: Myths, Machines, and Ancient Dreams of Technology* (Princeton, NJ: Princeton University Press, 2018).

32. Donna J. Haraway, "The High Cost of Information in Post World War II Evolutionary v Biology: Ergonomics, Semiotics, and the Sociobiology of Communications Systems," in *Philosophical Forum:: Sociobiology*, XIII, nos. 2/3, *The Debate Evolves* (Hoboken, NJ: Wiley-Blackwell, 1981–1982), 244–78.

33. Daniel S. Lehrman, "A Critique of Konrad Lorenz's Theory of Instinctive Behavior," *Quarterly Review of Biology* 28, no. 4 (1953): 337–63, DOI: 10.1086/399858.

34. Aubrey Manning and Marion Stamp Dawkins, *An Introduction to Animal Behaviour* (Cambridge: Cambridge University Press, 1992), 19.

35. Jonathan Silverton and Deborah M. Gordon, "A Framework for the Analysis of Plant Behavior," *Annual Review of Ecology and Systematics* 20, no. 1 (1989): 349–66, DOI: 10.1146/annurev.es.20.110189.002025; Anthony Trewavas, "Plant Intelligence," *Naturwissenschaften* 92, no. 9 (2005): 401–13, https://doi.org/10.1007/s00114-005-0014-9.

36. Krista K. Ingram, Deborah M. Gordon, Daniel A. Friedman, Michael K. Greene, John Kahler, and Swetha Peteru, "Context-Dependent Expression of the *Foraging* Gene in Field Colonies of Ants: The Interacting Roles of Age, Environment, and Task," *Proceedings of the Royal Society B: Biological Sciences* 283, no. 1837 (2016), https://doi.org/10.1098/rspb.2016.0841.

37. Greene and Gordon, "Cuticular Hydrocarbons Inform Task Decisions."

38. Michael J. Greene and Deborah M. Gordon, "How Patrollers Set Foraging Direction in Harvester Ants," *American Naturalist* 170 (2007): 943–48.

39. Jacob D. Davidson, Roxana P. Auraco-Aliaga, Sam Crow, Deborah M. Gordon, and Mark S. Goldman, "Effect of Interactions between Harvester Ants on Forager Decisions," *Frontiers in Ecology and Evolution* 4 (October 5, 2016), DOI: 10.3389/fevo.2016.00115.

40. Jacob D. Davidson and Deborah M. Gordon, "Spatial Organization and Interactions of Harvester Ants during Foraging Activity," *Journal of the Royal Society Interface* 14, no. 135 (2017), https://doi.org/10.1098/rsif.2017.0413; Cameron Musco, Hsin-H Su, and Nancy A. Lynch, "Ant-Inspired Density Estimation via Random Walks," *Proceedings of the National Academy of Science* 114, no. 40 (2017): 10534–41, https://doi.org/10.1073/pnas.1706439114.

41. Noa Pinter-Wollman, Ashwin Bala, Andrew Merrell, Jovel Queirolo, Martin C. Stumpe, Susan Holmes, and Deborah M. Gordon, "Harvester Ants Use Interactions to Regulate Forager Activation and Availability," *Animal Behaviour* 86, no. 1 (2013): 197–207, https://doi.org/10.1016/j.anbehav.2013.05.012; Evlyn Pless, Jovel Queirolo, Noa Pinter-Wollman, Sam Crow, Kelsey Allen, Maya B. Mathur, and Deborah M. Gordon, "Interactions Increase Forager Availability and Activity in Harvester Ants," *PLoS ONE* 10, no. 11 (2015): e0141971, DOI: 10.1371/journal.pone.0141971.

42. Renato Pagliara, Deborah M. Gordon, and Naomi Ehrich Leonard, "Regulation of Harvester Ant Foraging as a Closed-Loop Excitable System," *PLoS Computational Biology* 14, no. 12 (2018): e1006200, DOI: 10.1371/journal.pcbi.1006200.

43. Nicole Nova, Renato Pagliara, and Deborah M. Gordon, "Individual Variation Does Not Regulate Forager Response to Humidity in Harvester Ants," *Frontiers in Ecology and Evolution* 9 (2022), https://doi.org/10.3389/fevo.2021.756204.

44. Friedman et al., "The Role of Dopamine in the Collective Regulation of Foraging in Harvester Ants."

45. Blair D. Beverly, Helen McLendon, S. Nacu, Susan P. Holmes, and Deborah M. Gordon, "How Site Fidelity Leads to Individual Differences in the Foraging Activity of Harvester Ants," *Behavioral Ecology* 20, no. 3 (2009): 633–38, DOI:10.1093/beheco/arp041.

46. Gordon, "The Regulation of Foraging Activity in Red Harvester Ant Colonies"; Shelby J. Sturgis, Michael J. Greene, and Deborah M. Gordon, "Hydrocarbons on Harvester Ant (*Pogonomyrmex barbatus*) Middens Guide Foragers to the Nest," *Journal of Chemical Ecology* 37, no. 5 (2011): 514–24, DOI: 10.1007/s10886-011-9947-y.

47. See, for example, Thomas Pradeu, "The Organism in Developmental Systems Theory," *Biological Theory* 5, no. 3 (2010): 217–22, DOI: 10.1162/BIOT_a_00042.

Chapter 9. From the Collective to the Individual

1. Elisabeth A. Lloyd, "Units and Levels of Selection," in *Stanford Encyclopedia of Philosophy*, August 22, 2005, substantially revised April 14, 2017, https://plato.stanford.edu/entries/selection-units/.

2. Elisabeth A. Lloyd and Michael J. Wade, "Criteria for Holobionts from Community Genetics," *Biological Theory* 14, no. 4 (2019): 151–70, https://doi.org/10.1007/s13752-019-00322-w; Michael J. Wade, "A Critical Review of the Models of Group Selection," *Quarterly Review of Biology* 53, no. 2 (1978): 101–14, DOI: 10.1086/410450.

3. Elliott Sober and David Sloan Wilson, *Unto Others: The Evolution and Psychology of Unselfish Behavior* (Cambridge, MA: Harvard University Press, 1988); Samir Okasha, *Evolution and the Levels of Selection* (Oxford: Clarendon Press, 2006); Peter Godfrey-Smith, *Darwinian Populations and Natural Selection* (Oxford: Oxford University Press, 2009).

4. Ryan K. Balot, *Greek Political Thought* (Malden, MA: Blackwell Publishing, 2006).

5. Richard Tuck, ed., *Hobbes: Leviathan* (Cambridge: Cambridge University Press, 1996).

6. Adam Smith, *An Inquiry into the Nature and Causes of the Wealth of Nations* (London: Strahan and Cadell, 1776).

7. Garrett Hardin, "Tragedy of the Commons," *Science* 162, no. 3859 (1968): 1243–48, DOI: 10.1126/science.162.3859.1243.

8. Pyotr Kropotkin, *Mutual Aid: A Factor of Evolution* (Boston: Porter Sargent Publishers, 1902).

9. W. C. Allee, *The Social Life of Animals* (New York: W. W. Norton & Co., 1938).

10. Mark E. Borrello, "'Mutual Aid' and 'Animal Dispersion': An Historical Analysis of Alternatives to Darwin," *Perspectives in Biology and Medicine* 47, no. 1 (2004): 15–31, DOI: 10.1353/pbm.2004.0003.

11. George Williams, *Adaptation and Natural Selection* (Princeton, NJ: Princeton University Press, 1966).

12. Richard Dawkins, *The Selfish Gene* (Oxford: Oxford University Press, 1976).

13. Patrick Bateson, "The Nest's Tale: A Reply to Richard Dawkins," *Biology and Philosophy* 21 (2006): 553–58.

14. Donna J. Haraway, *Primate Visions: Gender, Race, and Nature in the World of Modern Science* (New York: Routledge, 1989).

15. N. B. Davies, *Dunnock Behaviour and Social Evolution* (Oxford: Oxford University Press, 1992), 1.

16. John H. Fanshawe and Clare D. Fitzgibbon, "Factors Influencing the Hunting Success of an African Wild Dog Pack," *Animal Behaviour* 45, no. 3 (1993): 479–90, https://doi.org/10.1006/anbe.1993.1059.

17. Mancur Olson, *The Logic of Collective Action* (Cambridge, MA: Harvard University Press, 1965).

18. R. J. Groom, K. Lannas, and Craig Ryan Jackson, "The Impact of Lions on the Demography and Ecology of Endangered African Wild Dogs," *Animal Conservation* 20, no. 4 (2017): 382–90, https://doi.org/10.1111/acv.12328.

19. See, for example, Federica Carugati and Margaret Levi, *A Moral Political Economy: Present, Past, and Future*, in *Elements in Political Economy*, edited by David Stavage (Cambridge: Cambridge University Press, 2021).

20. Williams, *Adaptation and Natural Selection*; Stephen J. Gould and Richard C. Lewontin, "The Spandrels of San Marco and the Panglossian Paradigm: A Critique of the Adaptationist Programme," *Proceedings of the Royal Society B: Biological Sciences* 205, no. 1161 (1979): 581–98, https://doi.org/10.1098/rspb.1979.0086.

21. Richard C. Lewontin, "The Wars over Evolution," *New York Review of Books*, October 20 2005.

22. *Plato: Euthyphro, Apology, Crito, Phaedo*, translated by Chris Emlyn-Jones and William Preddy, Loeb Classical Library 36 (Cambridge, MA: Harvard University Press, 2017).

23. William D. Hamilton, "The Genetical Evolution of Social Behaviour," *Journal of Theoretical Biology* 7, no. 1 (1964): 1–16.

24. See, for example, Chris R. Smith, Sara Helms Cahan, Carsten Kemena, Seán G. Brady, Wei Yang, Erich Bornberg-Bauer, Ti Eriksson, Juergen Gadau, Martin Helmkampf, Dietrich Gotzek, Misato Okamoto Miyakawa, Andrew V. Suarez, and Alexander Mikheyev, "How Do Genomes Create Novel Phenotypes? Insights from the Loss of the Worker Caste in Ant Social Parasites," *Molecular Biology and Evolution* 2, no. 11 (2015): 2919–2931, https://doi.org/10.1093/molbev/msv165; West-Eberhard, *Developmental Plasticity and Evolution*.

25. Walter D. Koenig, Frank A. Pitelka, William J. Carmen, Ronald L. Mumme, and Mark T. Stanback, "The Evolution of Delayed Dispersal in Cooperative Breeders," *Quarterly Review of Biology* 67 (1992): 111–50, https://doi.org/10.1086/417552.

26. See, for example, Charles R. Brown, Mary Bomberger Brown, Erin A. Roche, Valerie A. O'Brien, and Catherine E. Page, "Fluctuating Survival Selection Explains Variation in Avian Group Size," *Proceedings of the National Academy of Science* 113, no. 18 (2016): 5113–18, https://www.jstor.org/stable/26469507.

27. Dustin R. Rubenstein and Irby R. Lovette, "Temporal Environmental Variability Drives the Evolution of Cooperative Breeding in Birds," *Current Biology* 17, no. 16 (2007): 1414–19, DOI: 10.1016/j.cub.2007.07.032; Sarah Guindre-Parker and Dustin R. Rubenstein, "Survival Benefits of Group Living in a Fluctuating Environment," *American Naturalist* 195, no. 6 (2020): 1027–36, DOI: 10.1086/708496.

28. Thanks to Judy Bronstein for pointing this out.

29. Kaitlyn A. Mathis and Judith L. Bronstein, "Our Current Understanding of Commensalism," *Annual Review of Ecology, Evolution, and Systematics* 51 (2020): 167–89, https://doi.org/10.1146/annurev-ecolsys-011720-040844; Judith L. Bronstein, "The Study of Mutualism," in *Mutualism*, edited by Judith L. Bronstein (Oxford: Oxford University Press, 2015), 3–19.

30. Elizabeth G. Pringle, Erol Akçay, Ted K. Raab, Rodolfo Dirzo, and Deborah M. Gordon, "Water Stress Strengthens Mutualism among Ants, Plants, and Scale Insects," *PloS Biology* 11, no. 11 (2013): e1001705, DOI: 10.1371/journal.pbio.1001705.

31. Megan E. Frederickson, "Rethinking Mutualism Stability: Cheaters and the Evolution of Sanctions," *Quarterly Review of Biology* 88, no. 4 (2013): 269–95, DOI: 10.1086/673757.

32. Emilie C. Snell-Rood, "Evolutionary Causes and Consequences of Behavioral Plasticity," *Animal Behaviour* 85, no. 5 (2013): 1004–11, https://doi.org/10.1016/j.anbehav.2012.12.031; Niels J. Dingemanse, Anahita J. N. Kazem, Denis Réale, and Jonathan Wright, "Behavioural Reaction Norms: Animal Personality Meets Individual Plasticity," *Trends in Ecology and Evolution* 25, no. 2 (2010): 81–89, https://doi.org/10.1016/j.tree.2009.07.013; Renée A Duckworth, "The Role of Behavior in Evolution: A Search for Mechanism," *Evolutionary Ecology* 23, no. 4 (2009): 513–31, https://doi.org/10.1007/s10682-008-9252-6.

33. Samir Okasha, "Cancer and the Levels of Selection," *British Journal for the Philosophy of Science* (June 2021), https://doi.org/10.1086/716178.

34. Deborah M. Gordon, "The Rewards of Restraint in the Collective Regulation of Foraging by Harvester Ant Colonies," *Nature* 498, no. 7452 (2013): 91–93, DOI: 10.1038/nature12137.

35. Daniel Ari Friedman, Ryan Alexander York, Austin Travis Hilliard, and Deborah M. Gordon, "Gene Expression Variation in the Brains of Harvester Ant Foragers Is Associated with Collective Behavior," *Communications Biology* 3, no. 1 (2020): 100, DOI: 10.1038/s42003-020-0813-8.

36. Mekala Sundaram, Erik Steiner, and Deborah M. Gordon, "Rainfall, Neighbors and Foraging: The Dynamics of a Population of Harvester Ant Colonies 1988–2019," *Ecological Monographs* 92, no. 2 (2022): e1503, https://doi.org/10.1002/ecm.1503.

INDEX

Page numbers in *italics* indicate figures and tables.